BIOREMEDIATION

Green Solution of Environmental Pollutants

BIOREMEDIATION

Green Solution of Environmental Pollutants

Prof. Ram Chandra

Dr. Kshitij Singh

Department of Environmental Microbiology
Babasaheb Bhimrao Ambedkar University
(A Central University), Lucknow

CRC Press
Taylor & Francis Group
Boca Raton London New York

CRC Press is an imprint of the
Taylor & Francis Group, an **informa** business

-EPH-

Elite Publishing House

First edition published 2026
by CRC Press
4 Park Square, Milton Park, Abingdon, Oxon, OX14 4RN

and by CRC Press
2385 NW Executive Center Drive, Suite 320, Boca Raton FL 33431

CRC Press is an imprint of Informa UK Limited

British Library Cataloguing-in-Publication Data
A catalogue record for this book is available from the British Library

Print edition not for sale in India.

ISBN: 9781041331162 (hbk)
ISBN: 9781041333876 (pbk)
ISBN: 9781003787549 (ebk)

DOI: 10.1201/9781003787549

Typeset in A Caslon Pro
by Elite Publishing House

-EPH-

Contents

Preface

In an era marked by environmental concerns and the pursuit of sustainable practices, the field of bioremediation has emerged as a beacon of hope. "Bioremediation: Green Solution of Environmental Pollutants" is a comprehensive journey through the world of this innovative and eco-friendly approach to mitigate environmental pollutants and restore the affected ecosystem. Present book is a testament of green technology for sustainable development. Furthermore, the aim of this book is to provide simple and clear concept to understand the principle of bioremediation and biodegradation. The various chapters has emphasized on core concepts and principles that underpin bioremediation. Readers will gain an understanding of basic knowledge how bioremediation harnesses the power of plants and microbes to combat pollutants for sustainable development. Furthermore, book also has emphasized fundamental process of breaking down contaminants of environment. We have unraveled the intricacies of biodegradation and its integral role in bioremediation. It has focused light on the difference between biodegradation and bioremediation. The chapters have also described detail process of in-situ and ex-situ process. As plants are essential component of ecosystem for maintaining the life on earth therefore the concept and mechanism of phytoremediation has been well described to understand the broader range of readers. The book continues to diversify, delving into the bioremediation of lignocellulosic waste, emphasizing the importance of managing diverse biomass materials. Since the heavy metal contamination in environment has also posed threat to human health due to accumulation in food chain and other step of life. Therefore, the significant light has been focused on heavy metal toxicity in the environment and its bioremediation process. The biodegradation of xenobiotic compounds are also elaborated in systematic manners so that the knowledge of book may be as foundation stone to readers for development of innovative technology. The chapters has also elaborated innovative field of nanobioremediation, which employs nanomaterials to augment the remediation process. At the last the book has described the various factors that influence the success and efficiency of bioremediation efforts, highlighting the complexity of environmental remediation.

"Bioremediation: Green Solution of Environmental Pollutants" is designed to serve as a comprehensive reference for students, researchers, and practitioners engaged in the quest for cleaner, healthier environments. It is our hope that this book will empower readers with the knowledge and tools they need to make a positive impact in the field of bioremediation. Together, we embark on a journey to restore the natural world and create a more sustainable future.

Authors

Preface

About the Authors

Dr. Ram Chandra is currently Professor in the Department of Environmental Microbiology at Babasaheb Bhimrao Ambedkar University. He is among the senior professors of the university. He has completed his educational qualification from Banaras Hindu University (B.H.U.) and thereafter he joined as scientist B at CSIR-Indian Institute of Toxicology Research (IITR) in September, 1989. He has served to CSIR-IITR more than 23 years from Scientist (Sci. B) to Sr. Principal Scientist (Sci. F). Subsequently, he joined as Professor and Head in department of Environmental Microbiology at B. B. Ambedkar Central University in July 2011. His major contribution in area of Bioremediation of distillery and lignocellulosic waste generated from pulp paper industry which is a major threat to environment due to complex pollutants discharge from these industry. He has also praiseworthy leading work on bacterial assisted phyto-nanobioremediation of industries. He has published 240 research publication in peer reviewed journals and 07 books from CRC Press, U.S.A. His citation is: 10084, h-index: 58 and i10-index: 129. He is fellow of Biotech Research Society (BRSI), India and Association of Microbiologists of India (AMI), Fellow of Academy of Environmental Biology (AEB), Member of National Academy of Sciences, India (NASI).

He has visited Japan, Spain, Italy, & England, UK, for his scientific collaboration and invited talk based on his expertise. He has been honored with various awards also. He is also among the world top 2% ranking Scientist as per Stanford University Research, USA.

Dr. Kshitij Singh is currently a Guest Faculty member in the Department of Environmental Microbiology at Babasaheb Bhimrao Ambedkar University in Lucknow, where he also earned his doctorate in the subject of Environmental Microbiology. He completed his Master's degree in Plant Biotechnology at Banaras Hindu University in Varanasi. Additionally, he holds the position of Joint Secretary in the Microbiological Association for Science and Technology Development (MASTD), a society registered under the Society Registration Act XXVI, 1860, in Uttar Pradesh.

Dr. Singh has an impressive publication record, having authored several reviews and research articles in high-impact journals. He is also the holder of an international patent. Furthermore, he serves as a reviewer for several Springer and Elsevier journals.

Chapter - 1

Concept, Principle and Mechanism of Bioremediation

1.1. Introduction

Micro-organisms are well known for their ability to break down a huge range of organic compounds and absorb inorganic substances as their nutrient for growth, due to their broad range adaptation in environment. The term Bioremediation is defined as "Use of living organisms to clean up or remove pollutants from soil, water, or wastewater into harmless state or less toxic form (Pande et al., 2020). In bioremediation the process is primarily mediated by broad range of microorganism in the environment i.e. bacterial, fungi, actinomycetes, algae, cyanobacteria etc.

Despite their potential to remediate and adapt to polluted environments, it's important to note that not all pollutants can be easily handled by microbes, and the introduction of certain microbes into ecosystems should be carefully considered to prevent unintended consequences. There are various sources of environmental pollutants from different anthropogenic activities including the industrial waste (Ukaogo et al., 2020). In the industrial waste there are several hazardous and complex waste which do not degrade easily. For example, pesticides and insecticides, agricultural runoff, pharmaceutical waste, industrial chemicals and heavy metals, electronic waste (E-waste). The cumulative impact of these hazardous industrial wastes necessitates robust regulations, sustainable practices, and technological innovations to mitigate their environmental footprint and safeguard the well-being of ecosystems and communities.

At the initial stage of bioremediation microorganism interact with pollutants in favourable environment either by releasing some enzymes or absorbing the contaminant directly to their cell cytoplasm. Generally secreted enzymes from microorganism interact with those pollutants which cannot enter or absorb inside the

DOI: 10.1201/9781003787549-1

microbial cell they are broken down in .its component which can directly interact to the cellular signal for their absorption this will result for reduction of pollutant present at any site (Narayanan et al., 2023). In the process of bioremediation microorganism converts the various categories of pollutants into their constituent unit as end products, eg. Carbon (C), Nitrogen (N), Hydrogen (H), Oxygen (O) or CO_2 which are the major component of various substances. The enzymes present inside the microbial cellular system are known as constitutive enzymes. Some enzymes are induced by organism in presence of pollutants; these enzymes are known as inducible enzyme. In the Earth's biosphere, microorganisms grow in the widest range of habitats. They grow in soil, water, plants, animals, deep sea, and freezing ice environment. Their absolute numbers and their appetite for a wide range of chemicals make microorganisms the perfect candidate for acting as our environmental caretakers.

Bacteria, fungi, and plants are used in bioremediation to break down, remove, change, immobilize, or detoxify different chemicals and physical wastes from the environment. Through their enzymatic pathways, microorganisms play a role as biocatalysts, facilitating the advancement of biochemical reactions that break down the targeted pollutant. Only when they have access to a variety of resources and substances that enable them to produce energy and nutrition to create more cells do microorganisms take action against the pollutants. The chemical makeup and quantity of contaminants, the physicochemical properties of the environment, and their accessibility to microorganisms are just a few of the variables that affect bioremediation's effectiveness. Bacteria and contaminants do not come into contact with one another, which affects the rate of degradation. Furthermore, contaminants and microorganisms are not evenly distributed across the environment. Numerous elements make bioremediation process optimization and control a complex system. The presence of a microbial population that can break down the pollutants, the availability of contaminants to the microbial population, and environmental factors (such as soil type, temperature, pH, the presence of oxygen or other electron acceptors, and nutrients) are all included here (Abatenh et al., 2017).

General process of bioremediation

Process typically works:

Step 1: Environment: Bioremediation takes place in a contaminated environment, such as soil, water, or air, where pollutants are present. These pollutants can be organic compounds, such as petroleum hydrocarbons, solvents, or pesticides, or inorganic compounds like heavy metals.

Step 2: Microorganisms: Microorganisms play a crucial role in bioremediation. Certain bacteria, fungi, and other microorganisms have the ability to break down or transform pollutants into less harmful substances. These microorganisms are naturally present in the environment or can be introduced or enhanced through various methods. If microorganisms are already present in the culture medium, they may not go through a significant lag phase because they are adapted to the environment. On the other hand, microorganisms introduced later might experience a lag phase as they adjust to the new conditions before entering the log phase of active growth. Additionally, the overall growth curve might include additional phases such as stationary phase (growth rate slows down) and death phase (cell death exceeds cell division) depending on the specific conditions of the culture.

Step 3: Pollutants: The pollutants present in the contaminated environment serve as a food source or energy substrate for the microorganisms. They can metabolize or convert the pollutants into less toxic compounds through various biochemical pathways.

Step 4: Interactions: The microorganisms interact with the pollutants and the environment in a complex manner. Factors such as temperature, pH, oxygen availability, nutrient levels, and the presence of other chemicals can influence the biodegradation process. It is essential to optimize these environmental conditions to promote the growth and activity of the microorganisms.

Step 5: Biodegradation: As the microorganisms metabolize the pollutants, they break them down into simpler compounds, which may be further transformed or assimilated by other microorganisms in the ecosystem. Ultimately, the goal is to convert the pollutants into non-toxic substances like water, carbon dioxide, and harmless minerals.

Step 6: Monitoring and Control: Throughout the bioremediation process, it is important to monitor the progress and effectiveness of the treatment. Factors such as pollutant concentration, microbial activity, and environmental conditions are assessed to ensure that the remediation is proceeding as intended. Adjustments may be made if necessary to optimize the process.

Bioremediation offers several advantages over other remediation methods, such as being cost-effective, environmentally friendly, and applicable to a wide range of contaminants. However, the success of bioremediation depends on various factors, including the specific pollutants, the site conditions, selection and performance of appropriate microorganisms.

1.2. Concept of Bioremediation

Etymologically, the concept of bioremediation is derived from two words, "bios" meaning life (this refers to living organisms) and "to remediate" that means to solve a problem. "Bioremediate" thus means to use biological organisms to solve an environmental problem such as contaminated soil or groundwater. Thus, Bioremediation can be defined as the use of living microorganisms to degrade environmental pollutants or to prevent pollution. In other words, it is a technology for removing pollutants from the environment thus restoring the original natural surroundings and preventing further pollution (Wang et. al., 2021). The concept of bioremediation was first used on a large scale in 1972 for the cleaning of Sun Oil pipeline spill at Ambler, Pennsylvania. However, in laboratory-scale, George M. Robinson was the first to recognize this process during experiments on sewage and oil treatment. The fundamental concept of bioremediation by microbes involves using microorganisms, such as bacteria, fungi, and algae, to degrade, transform, or remove contaminants from polluted environments. Microbes play a central role in bioremediation due to their unique ability to metabolize various pollutants, breaking them down into simpler and less harmful substances.

It is pertinent to note that bioremediation and biodegradation do not possess the same meaning. Bioremediation as an environmental clean-up strategy may include biodegradation as only one of the mechanisms involved or applied in the bioremediation process. Only some of contaminants are biodegradable, and only some of microorganisms can degrade a fraction of contaminants. Despite the fact that microorganisms have been used for the treatment and transformation of waste products for at least a century, bioremediation is considered as a developing technology for eco-friendly decontamination of polluted environment.

To put this into context, in theory there are enough bioremediants in nature that can be applied against a broad range of pollutants and bioremediation can be considered as a useful technique for the complete destruction of a wide variety of contaminants. Many compounds that are legally regarded as detrimental and dangerous can thus be bio-transformed to harmless products. This eliminates the chance of future liability associated with treatment and disposal of contaminated material. Instead of transferring contaminants from one environmental medium to another, for example, from land to water or air, the complete destruction of target pollutants is possible. Bioremediation can often be carried out on site, often without causing a major disruption of normal activities. This also eliminates the need to transport quantities of waste off site and the potential threats to human health and the environment that can arise during transportation.

However, notwithstanding the numerous advantages, a bioremediation technology suffers from primary major drawbacks. One is that only a few bacteria and fungi act on a broad range of organic compounds. In other words, no organism has been known to be omnivorous enough to destroy a large percentage of the natural chemicals.

Notwithstanding, this drawback may be bypassed through screenings for the discovery of new microbial species and detection of the potential of microorganisms in one hand, and the synchronized or successive application of the microorganisms that complete the bioactivity of each other. Another drawback to bioremediation is that it takes a long period of time to act and impose its effect. There are some solutions to get rid of such a limitation. Genetic manipulation techniques have led to an invaluable opportunity to obtain new strains of enhanced or new bioremediation activity. Another solution may be the addition of some enhancers in an environment or in a formulation to biochemically fortify certain bioremediation pathway (s). The third solution is the synchronous use of one or more microorganisms that directly and/ or indirectly increase the bioactivity of a bioremediant. Increase of the bioremediant population is the fourth and the simplest solution.

Bioremediation is limited to those compounds that are biodegradable. Not all compounds are susceptible to rapid and complete degradation, and there are some biodegradable products that may be more persistent or toxic than the parent compounds. Biological processes are often highly specific. Important site factors required for successful bioremediation include the presence of metabolically capable microbial populations, suitable environmental growth conditions, and appropriate levels of nutrients and contaminants. It is difficult to extrapolate from bench and pilot-scale studies to full-scale field operations. Research is needed to develop and engineer bioremediation technologies that are appropriate for sites with complex mixtures of contaminants that are not evenly dispersed in the environment. Generally, contaminants may be present as solids, liquids, and gases. But irrespective of the contaminant, bioremediation often takes longer than other treatment options, such as excavation and removal of soil or incineration. Since biological process remain very specific which require the optimum condition to initiate all inside molecular enzymatic action of microbes which may be regulated by presence of required size of microbial population (Inoculum size). But in bioremediation and biodegradation the compound has to be broken into its elemental end product which may be ultimately taken by microbes as their nutrient and ecosystem restored to its position. Therefore, to assess the bioremediation process their microbial population or involved enzyme may be evaluated.

The key elements of the concept include:

i. *Biodegradation*: Microbes possess specific enzymes that allow them to break down complex organic compounds found in contaminants, such as hydrocarbons, pesticides, and solvents. These enzymes act as biological catalysts, facilitating chemical reactions that transform the pollutants into simpler and less toxic forms.

Microorganisms use the contaminants as a source of energy or nutrients for their growth and metabolic processes. For example, certain bacteria have enzymes known as hydroxylases that can add hydroxyl groups to hydrocarbons, making them more water-soluble and easier to degrade. Other microorganisms possess enzymes like dehalogenases that can remove halogen atoms from halogenated organic compounds, reducing their toxicity. By utilizing these pollutants as a source of energy or nutrients, microorganisms convert them into less toxic forms, ultimately reducing their toxicity and environmental impact. By taking advantage of the natural abilities of these microorganisms, bioremediation offers a sustainable and eco-friendly approach to address pollution and restore environmental health. However, the success of bioremediation depends on various factors, including the type and concentration of contaminants, environmental conditions, and the availability of suitable microorganisms for the specific cleanup task.

ii. *Aerobic and Anaerobic Processes*: Some microorganisms work best in the presence of oxygen (aerobic conditions), while others thrive in oxygen-depleted environments (anaerobic conditions). This diversity allows microbes to address pollution in various environmental settings, such as surface soils and groundwater.

- Aerobic Conditions: In aerobic environments where oxygen is present, certain microorganisms, known as aerobic microbes, thrive and carry out biodegradation processes. Aerobic bioremediation is the most common form of microbial cleanup. These microbes use oxygen as a terminal electron acceptor during respiration, which allows them to extract energy from organic compounds present in pollutants. Aerobic biodegradation is generally faster and more efficient compared to anaerobic processes. It is particularly effective in breaking down a wide range of contaminants, including many hydrocarbons and some organic chemicals.

- Anaerobic Conditions: In oxygen-depleted environments, certain microorganisms, known as anaerobic microbes, are adapted to survive and function without oxygen. Anaerobic bioremediation occurs when these microbes utilize other substances, such as nitrates, sulfates, or even the contaminants themselves, as electron acceptors in their metabolic processes. Anaerobic biodegradation tends to be slower and less efficient than aerobic biodegradation, but it is effective for certain types of pollutants that are more resistant to aerobic degradation. Examples of pollutants that can be addressed by anaerobic bioremediation include chlorinated solvents and some highly reduced compounds.

iii. Nutrient Requirements: Microbial bioremediation often involves providing essential nutrients like nitrogen, phosphorus, and trace elements to support the growth and activity of pollutant-degrading microorganisms (Filler et. al., 2008). This process, known as biostimulation, enhances microbial activity and accelerates biodegradation. Therefore, providing essential nutrients becomes crucial to support the growth and activity of these pollutant-degrading microorganisms. The major elements required for microorganism with their source and function is shown in Table 1.1.

The key nutrients typically provided in microbial bioremediation are:

» Nitrogen: Nitrogen is an essential component of proteins and enzymes required for the microorganisms' metabolism. Often, nitrogen is provided in the form of ammonium or nitrate salts.

» Phosphorus: Phosphorus is a critical element for energy transfer and cell growth. It is supplied in the form of phosphate compounds.

» Trace Elements: Microorganisms require trace elements such as iron, manganese, copper, zinc, and molybdenum as cofactors for various enzymes involved in pollutant degradation.

Table 1.1. Major elements, their sources and functions in bacterial cells.

Element	% of dry weight	Source	Function
Carbon	50	organic compounds or CO_2	Main constituent of cellular material
Oxygen	20	H_2O, organic compounds, CO_2, and O_2	Constituent of cell material and cell water; O_2 is electron acceptor in aerobic respiration

Nitrogen	14	NH_3, NO_3, organic compounds, N_2	Constituent of amino acids, nucleic acids nucleotides, and coenzymes
Hydrogen	8	H_2O, organic compounds, H_2	Main constituent of organic compounds and cell water
Phosphorus	3	inorganic phosphates (PO_4)	Constituent of nucleic acids, nucleotides, phospholipids, LPS, teichoic acids
Sulfur	1	SO_4, H_2S, SO, organic sulfur compounds	Constituent of cysteine, methionine, glutathione, several coenzymes
Potassium	1	Potassium salts	Main cellular inorganic cation and cofactor for certain enzymes
Magnesium	0.5	Magnesium salts	Inorganic cellular cation, cofactor for certain enzymatic reactions
Calcium	0.5	Calcium salts	Inorganic cellular cation, cofactor for certain enzymes and a component of endospores
Iron	0.2	Iron salts	Component of cytochromes and certain nonheme iron-proteins and a cofactor for some enzymatic reactions

iv. *Bioaugmentation*: In certain cases, specific pollutant-degrading microorganisms may be introduced into the contaminated site to enhance the bioremediation process. These microorganisms can be native to the environment or from external sources. This approach, known as bioaugmentation, helps to boost the overall microbial population and degradation efficiency.

The role of bioaugmentation in bioremediation is to (Muter, 2023):

» Increase degradation capabilities: In some cases, the native microorganisms in the contaminated site might not be efficient enough to degrade certain pollutants or may be inhibited by the presence of toxic substances. By introducing highly effective and adapted microorganisms through bioaugmentation, the degradation capabilities of the microbial community can be enhanced, leading to more efficient and rapid bioremediation.

» Targeted approach: Bioaugmentation allows researchers and environmental engineers to select specific microbial strains that are known to possess the desired enzymatic machinery for breaking down particular contaminants.

This targeted approach ensures that the introduced microorganisms can directly tackle the specific pollutants of concern.

- Adaptability: In situations where the environmental conditions are harsh, such as high levels of toxicity or extreme pH levels, native microorganisms might struggle to survive and function optimally. Bioaugmentation can introduce microorganisms that are better adapted to such conditions, thereby improving the chances of successful bioremediation.

- Site-specific solutions: Different contaminated sites can have unique pollutants and environmental conditions. Bioaugmentation provides the flexibility to tailor the bioremediation process to the specific needs of each site, as different microbial strains can be chosen and utilized accordingly.

- Accelerated bioremediation: By introducing specialized microorganisms that are highly efficient in breaking down the target pollutants, the overall bioremediation process can be accelerated, reducing the time required to clean up the contaminated site.

v. *Synergistic Interactions*: Microorganisms can work synergistically, cooperating with each other to degrade complex pollutants more effectively than individual microbes. This cooperative behavior can lead to increased pollutant degradation rates (Khanpour and Partovinia, 2021). These interactions involve the cooperation and coordination between different microorganisms, as well as between microorganisms and other factors in the environment. Some of the key synergistic interactions in bioremediation include:

- Co-metabolism: Co-metabolism occurs when one microorganism produces enzymes or metabolites that facilitate the breakdown of a pollutant, even though the pollutant itself is not used as a primary source of energy by the microorganism. Other microorganisms in the environment may benefit from these by-products, leading to increased pollutant degradation overall.

- Synergistic Microbial Consortia: Certain microorganisms may work more effectively together in a consortium than individually. For example, one microorganism may produce specific enzymes that break down complex pollutants, while another microorganism can use the resulting breakdown products as a source of energy, leading to a more efficient degradation process.

- Cross-feeding: Cross-feeding occurs when different microorganisms exchange metabolites or breakdown products with each other. In bioremediation, this can lead to the removal of contaminants through a network of interactions between various microbial species, where one species provides nutrients or intermediates required by another species for pollutant degradation.

» Nutrient Exchange: Some microorganisms may release excess nutrients into the environment, which can be utilized by other microorganisms for growth and activity. This exchange of nutrients supports a diverse microbial community and enhances the degradation potential of the overall system.

» Biofilm Formation: Microorganisms often form biofilms, which are organized communities of cells embedded in a matrix. Biofilms facilitate close interactions between different microorganisms and can enhance the degradation of pollutants by providing a protected and stable environment for microbial activities.

» Redox Reactions: In many bioremediation processes, redox reactions are involved in pollutant degradation. Different microorganisms may be responsible for different stages of the redox reactions, and their cooperative action is essential for the complete breakdown of the pollutant.

» Enhanced Abiotic Factors: Some microorganisms can produce extracellular substances, such as biosurfactants, that modify abiotic factors like soil or water properties. These modifications can enhance the availability of pollutants to other microorganisms, facilitating their degradation.

» Bioaugmentation and Natural Microbiota: Bioaugmentation, as mentioned earlier, involves introducing specialized microorganisms to the contaminated site. These introduced microorganisms can potentially interact with the natural microbiota, sharing traits and functions that improve the overall bioremediation process.

vi. Adaptability: Adaptability is a critical aspect of bioremediation, as it refers to the ability of microorganisms and the overall bioremediation process to adjust and respond to changes in the environmental conditions and the presence of contaminants (Panigrahi et al., 2019). The success of bioremediation depends on the ability of the microbial community to adapt to the specific challenges posed by the pollutants and the surrounding environment. Here's how adaptability plays a crucial role during bioremediation:

» Tolerance to Environmental Conditions: Microorganisms involved in bioremediation must be able to tolerate a wide range of environmental conditions, such as temperature, pH, salinity, and oxygen levels. Adaptability allows them to thrive and function effectively under varying conditions.

- Contaminant Specificity: Different contaminants require different enzymatic pathways for degradation. Microorganisms must adapt to the presence of specific pollutants and produce the appropriate enzymes needed for their breakdown.
- Co-metabolism: Some pollutants are not used as a primary source of energy by microorganisms but can still be degraded through co-metabolism. Microorganisms may adapt their metabolic pathways to utilize co-substrates and co-factors that facilitate the breakdown of contaminants.
- Genetic Adaptation: Over time, microbial populations can undergo genetic changes and adaptations through mutation and natural selection. Certain strains may evolve with enhanced capabilities for pollutant degradation, improving the overall bioremediation process.
- Microbial Community Dynamics: Bioremediation often involves diverse microbial communities. As conditions change, certain microorganisms may become more dominant, while others may decrease in abundance. This dynamic adaptation helps to maintain an efficient microbial consortium for pollutant degradation.
- Response to Inhibitors: Some contaminants or their breakdown products can be toxic to microorganisms. Adaptability allows microorganisms to develop mechanisms to resist or detoxify these inhibitors, ensuring their survival and continued degradation activities.
- Long-Term Remediation: Bioremediation processes can take considerable time to achieve complete cleanup. Adaptability allows the microbial community to persist over the long term, ensuring sustained degradation efforts.
- Bioaugmentation Success: In bioaugmentation, the success of introducing specialized microorganisms depends on their ability to adapt to the new environment and interact with the existing microbial community in a productive manner.
- Microbial Interactions: Adaptability enables microorganisms to establish positive interactions, such as cross-feeding and biofilm formation, that enhance the overall degradation capabilities of the microbial community.

- » Monitoring and Management: Understanding the adaptability of microorganisms during bioremediation helps in monitoring the progress of the cleanup process. If certain aspects of the process are not working as expected, adjustments can be made to optimize the remediation efforts.

vii. Monitoring and Optimization: Successful bioremediation projects require careful monitoring of microbial activity and the environmental conditions (Celin et al., 2020). By understanding the progress of biodegradation and adjusting factors like nutrient levels, pH, and oxygen availability, the process can be optimized for better results.

Monitoring:

- » Contaminant Levels: Regular monitoring of contaminant concentrations in the soil, water, or air is essential to assess the progress of bioremediation. This monitoring helps determine if pollutant levels are decreasing over time.
- » Microbial Activity: Monitoring microbial activity through various techniques, such as measuring metabolic by-products or tracking changes in microbial populations, provides insights into the effectiveness of the microbial community in degrading contaminants.
- » Environmental Conditions: Monitoring environmental factors, such as temperature, pH, oxygen levels, and nutrient availability, ensures that the conditions are within the optimal range for the growth and activity of the microorganisms.
- » Biodegradation Products: Monitoring the breakdown products of contaminants helps confirm that biodegradation is occurring and that potentially harmful intermediates are being transformed into less toxic compounds.
- » Toxicity Assessment: Assessing the toxicity of the treated environment is crucial to ensure that the bioremediation process is not creating more harmful compounds or negatively impacting the ecosystem.
- » Microbial Diversity: Monitoring the diversity of the microbial community allows researchers to identify shifts in population structures and potential changes in community dynamics during the bioremediation process.
- » Bioaugmentation Success: If bioaugmentation is used, monitoring the survival

and activity of the introduced microorganisms is essential to evaluate their impact on pollutant degradation.

Optimization:

- Bioaugmentation Strategies: If bioaugmentation is employed, optimizing the selection and introduction of specific microorganisms that are most effective at degrading the contaminants can enhance bioremediation efficiency.
- Nutrient Supply: Adjusting and optimizing nutrient supply (e.g., carbon, nitrogen, phosphorus) can stimulate microbial growth and activity, thereby improving pollutant degradation rates.
- Environmental Conditions: Controlling and optimizing environmental factors, such as temperature, pH, and oxygen levels, can create an ideal environment for the microbial community to thrive and function optimally.
- Biostimulation Techniques: Biostimulation involves providing external electron donors or acceptors to enhance microbial activity. Optimizing the application of biostimulation techniques can significantly improve pollutant degradation rates.
- Bioreactor Design: For ex-situ bioremediation, optimizing bioreactor design and operation parameters can ensure maximum contact between microorganisms and pollutants, leading to more efficient pollutant removal.
- Bioremediation Timeline: Based on the monitoring data, adjustments can be made to the bioremediation timeline, ensuring that the process is given sufficient time to achieve the desired cleanup goals.
- Risk Assessment: Continuous risk assessment throughout the bioremediation process helps identify potential adverse effects and allows for timely corrective actions.
- Remediation Endpoint: Deciding on the desired endpoint of bioremediation and optimizing the process to achieve that endpoint is essential to avoid under- or over-treatment of the contaminated site.

Depending on the specific contaminants and the site conditions, different bioremediation strategies may be employed, such as biostimulation (providing nutrients to enhance microbial activity), bioaugmentation (introducing specific pollutant-degrading microorganisms), phytoremediation (using plants to remediate the environment), and mycoremediation (using fungi for remediation purposes).

1.3. Principle of Bioremediation

The principle of bioremediation is based on the use of biological processes and living organisms to degrade, transform, or remove pollutants and contaminants from the environment. The goal of bioremediation is to harness the natural capabilities of microorganisms, plants, or other living organisms to clean up and restore polluted or contaminated sites (Tyagi and Kumar, 2021). Several key principles underpin the concept and application of bioremediation:

i. Use of Living Organisms: Bioremediation relies on the use of living organisms, such as bacteria, fungi, plants, and algae, to perform the remediation process. These organisms possess inherent abilities or can be engineered to break down or neutralize specific contaminants.

ii. Natural Processes: Bioremediation capitalizes on natural processes and ecological interactions in the environment. It harnesses the existing microbial communities and their metabolic activities to target and degrade pollutants, mimicking or enhancing natural degradation processes.

iii. Site-Specific Approaches: Bioremediation strategies are tailored to specific contaminated sites and the type of pollutants present. Different organisms and techniques may be applied depending on factors like the nature and concentration of contaminants, environmental conditions, and the desired outcome.

iv. Sustainability and Environmental Friendliness: Bioremediation is considered a more sustainable and environmentally friendly remediation option compared to some traditional methods like incineration or chemical treatments. It often results in less secondary pollution and damage to the ecosystem.

v. Biodegradation and Transformation: The primary mechanisms of bioremediation involve biodegradation and transformation. Biodegradation refers to the breakdown of pollutants into simpler and less harmful compounds by microorganisms. Transformation involves changing the chemical structure of contaminants to non-toxic forms.

vi. Stimulation and Augmentation: Bioremediation can be achieved through biostimulation or bioaugmentation. Biostimulation involves providing nutrients, oxygen, or other substances to enhance the activity of indigenous microorganisms, stimulating them to degrade contaminants more efficiently. Bioaugmentation, on the other hand, introduces specific microorganisms with enhanced degradation capabilities to the contaminated site.

vii. Monitoring and Control: Successful bioremediation requires careful monitoring and control of environmental conditions and the progress of the remediation process. Regular assessment of contaminant levels and the activity of the microorganisms involved helps optimize the remediation efficiency.

viii. Potential Limitations: Bioremediation may have limitations depending on the complexity and persistence of the contaminants, the availability of suitable microorganisms, and the environmental conditions. In some cases, multiple remediation strategies or complementary methods may be necessary.

The overall process is as follows:

i. ***Pollutant Solubility in Media:*** The first step involves understanding the nature of the pollutants and their solubility in the media (water, soil, etc.). Some pollutants may be soluble in water, while others may be more associated with solid particles. Pollutant solubility in the media is important for bioavailability to bacteria because it determines whether the pollutants are in a form that the microorganisms can access and metabolize. The bioavailability of a pollutant refers to its accessibility and suitability for uptake and use by living microorganisms.

ii. ***Bioavailability:*** Bioavailability refers to the accessibility of the pollutants to the microorganisms. The pollutants need to be in a form that bacteria can uptake and metabolize (Alori et al., 2022). This can involve physical and chemical transformations of the pollutants to make them more accessible to the bacteria.

More detailed explanation of bioavailability during bioremediation:

» ***Physical Form and Solubility:*** The physical form and solubility of a pollutant influence its bioavailability. Soluble pollutants are typically more bioavailable to microorganisms because they can be directly taken up from the surrounding environment. Insoluble pollutants may need to undergo physical or chemical transformations to become more bioavailable.

» ***Adsorption and Desorption:*** Pollutants may adsorb onto soil particles or other surfaces, reducing their availability to microorganisms. The process of desorption, where pollutants are released from these surfaces, can enhance their bioavailability. Microorganisms may produce extracellular substances or enzymes that aid in the desorption of pollutants, making them more accessible.

» ***Chemical Speciation:*** The chemical form of a pollutant can impact its bioavailability. Some pollutants may exist in different chemical species, and certain species may be more readily taken up by microorganisms. Understanding the chemical speciation of pollutants is crucial for assessing their bioavailability.

» ***Microbial Degradation and Transformation:*** Microorganisms produce enzymes that facilitate the breakdown of complex pollutants into simpler, more bioavailable forms. This process, often referred to as microbial degradation, is a key step in bioremediation. The transformation of pollutants into metabolites that are more easily assimilated by microorganisms enhances bioavailability.

» ***Nutrient Availability:*** Microorganisms involved in bioremediation require nutrients for their growth and metabolic activities. The availability of essential nutrients, such as nitrogen, phosphorus, and trace elements, can influence the bioavailability of pollutants. In some cases, adding nutrients to the environment can enhance microbial activity and pollutant degradation.

» ***Environmental Conditions:*** Factors such as temperature, pH, and oxygen availability affect microbial activity and, consequently, pollutant bioavailability. Optimal environmental conditions support the growth and metabolic activities of microorganisms, ensuring that they can effectively utilize and degrade pollutants.

» ***Competition for Substrates:*** In natural environments, microorganisms may compete for the same substrates or pollutants. The presence of a diverse microbial community can influence the bioavailability of pollutants as different species may have varying abilities to utilize specific compounds.

iii. Interaction: Once the pollutants are bioavailable, bacteria interact with them through processes like adsorption, absorption, and enzymatic reactions. Microorganisms produce extracellular enzymes to break down complex pollutants into simpler forms that can be easily assimilated.

Overview of the interaction processes that take place when pollutants are in a bioavailable form:

» ***Adsorption and Attachment:*** Microorganisms can adsorb or attach to the surface of bioavailable pollutants. This attachment is often facilitated by microbial surface structures and extracellular polymeric substances. Adsorption allows microorganisms to be in close proximity to the contaminants, promoting subsequent interactions.

» ***Enzymatic Degradation:*** Microorganisms produce specific enzymes that facilitate the breakdown of bioavailable pollutants into simpler, more manageable forms. Enzymatic degradation is a key process in bioremediation, as it transforms complex pollutants into metabolites that can be assimilated by microorganisms for energy and growth.

iv. ***Metabolism:*** Bacteria metabolize the pollutants using various metabolic pathways. This often involves the introduction of specific enzymes that catalyze the breakdown of pollutants into less toxic or non-toxic forms (Dangi et al., 2019). Common metabolic pathways include aerobic and anaerobic respiration, as well as fermentation.

Overview of the metabolism process:

» ***Intracellular Metabolism:*** Once inside the microbial cells, bioavailable pollutants are subjected to intracellular metabolic processes. Microorganisms utilize specific metabolic pathways to break down the pollutants into simpler compounds. The choice of metabolic pathway depends on the nature of the pollutants and the metabolic capabilities of the microorganisms involved.

» ***Enzymatic Reactions:*** Enzymes play a central role in the metabolic degradation of pollutants. Microorganisms produce enzymes that catalyze the step-by-step breakdown of complex pollutants into smaller, more manageable molecules. These enzymatic reactions often involve the cleavage of chemical bonds, resulting in the transformation of the pollutant.

» ***Aerobic and Anaerobic Respiration:*** Microorganisms may engage in aerobic or anaerobic respiration to derive energy from the metabolized pollutants. In aerobic respiration, microorganisms use oxygen as a terminal electron acceptor, while in anaerobic respiration, alternative electron acceptors, such as nitrate or sulfate, are utilized. These respiratory processes contribute to the energy production required for microbial activities.

v. ***Energy Production:*** During the metabolic processes, bacteria derive energy from the pollutants. This energy is essential for their growth and survival. The pollutants serve as a carbon and energy source for the bacteria, driving their metabolic activities.

Here's a closer look at how microbial energy production occurs after the metabolism of pollutants:

- *Respiratory Processes:* Microorganisms often utilize respiratory processes to extract energy from the metabolized pollutants. In aerobic respiration, the microbial cells use oxygen as the terminal electron acceptor, and energy is generated through the electron transport chain. In anaerobic respiration, microorganisms employ alternative electron acceptors, such as nitrate, sulfate, or carbon dioxide. These respiratory processes result in the production of adenosine triphosphate (ATP), the primary energy currency of cells.
- *ATP Synthesis:* ATP synthesis is a key outcome of microbial energy production. During respiratory processes, the electron transport chain pumps protons across the cell membrane, creating a proton gradient. The flow of protons back into the cell through ATP synthase complexes drives the synthesis of ATP from adenosine diphosphate (ADP) and inorganic phosphate (Pi).
- *Energy for Cellular Activities:* The ATP generated through microbial energy production serves as a direct source of energy for various cellular activities. This includes the biosynthesis of cellular components, maintenance of cellular integrity, and the transport of nutrients across cell membranes. The energy obtained from pollutant metabolism is crucial for sustaining the life cycle and activities of microorganisms involved in bioremediation.
- *Growth and Reproduction:* The energy derived from pollutant metabolism is used for microbial growth and reproduction. Microorganisms assimilate the carbon and nutrients obtained from the pollutants into their cellular biomass, leading to an increase in population size. This growth is essential for the effective remediation of contaminated environments.
- *Maintenance of Cellular Functions:* In addition to supporting growth, microbial energy is utilized for the maintenance of cellular functions. This includes the repair of cellular structures, the synthesis of enzymes and other proteins, and the regulation of metabolic pathways. The energy obtained from pollutant metabolism ensures the overall health and functionality of microbial cells.

vi. ***Biosynthesis:*** As bacteria break down pollutants, they incorporate the resulting breakdown products into their cellular structures. This process contributes to the growth and reproduction of the bacterial population. The incorporation of pollutant-derived compounds into bacterial biomass can lead to the reduction of pollutant concentrations in the environment.

It's important to note that the success of bacterial bioremediation depends on factors such as the specific pollutants involved, the types of bacteria selected for the process, environmental conditions (temperature, pH, etc.), and the overall system design. Additionally, monitoring and optimizing these processes are critical for ensuring the efficiency of bioremediation efforts.

Abiotic and Biotic factors for bioremediation

Bioremediation, the use of microorganisms to degrade or remove environmental contaminants, is influenced by a variety of biotic (living) and abiotic (non-living) factors that impact the effectiveness of the process (Mishra et al., 2021). Understanding and managing these factors is crucial for successful bioremediation efforts. Here are some key biotic and abiotic factors that play a role in bioremediation:

Biotic Factors:

i. Microbial Diversity: The presence and diversity of naturally occurring microorganisms in the contaminated site are critical for bioremediation. A diverse microbial community increases the chances of finding organisms capable of degrading different types of pollutants.

ii. Microbial Activity: The metabolic activity and growth rates of the microbial community determine the rate of pollutant degradation. Factors such as temperature, pH, and nutrient availability influence microbial activity.

iii. Adaptation and Acclimation: The ability of microorganisms to adapt and acclimate to the presence of pollutants is essential. Over time, microbial populations can evolve to become more effective at degrading specific contaminants.

iv. Competition and Predation: In the contaminated environment, microorganisms may compete for resources, including the pollutants themselves. Predatory microorganisms can also impact the overall microbial community dynamics.

v. Bioaugmentation: As mentioned earlier, the introduction of selected microorganisms through bioaugmentation can enhance bioremediation by providing specialized strains that are better suited for degrading specific pollutants.

Abiotic Factors:

i. Temperature: Temperature affects microbial activity, and different microorganisms have optimal temperature ranges for growth and degradation. Bioremediation can be accelerated by maintaining temperatures within the ideal range.

ii. pH: Microbial activity is highly dependent on pH, and different microorganisms have different pH tolerances. Contaminants can also alter the pH of the environment, affecting microbial function.

iii. Oxygen Availability: Oxygen is essential for many biodegradation processes. Aerobic bioremediation relies on oxygen, while anaerobic bioremediation processes occur in the absence of oxygen.

iv. Moisture: Adequate moisture levels are necessary to support microbial growth and maintain activity. Extreme dryness or waterlogging can limit microbial function.

v. Nutrient Availability: Microorganisms require nutrients (e.g., carbon, nitrogen, phosphorus) for growth and metabolism. The availability of these nutrients can influence microbial activity and pollutant degradation.

vi. Presence of Inhibitors: Some pollutants or their breakdown products may inhibit microbial activity or even be toxic to certain microorganisms, impacting the overall bioremediation process.

vii. Soil Texture and Composition: The physical characteristics of the soil, such as texture and mineral composition, can affect microbial mobility, colonization, and activity.

viii. Presence of Co-contaminants: Co-existing pollutants in the contaminated site can interact with the target pollutants and affect microbial degradation processes.

To achieve successful bioremediation, it's important to carefully assess and manage these biotic and abiotic factors. Site-specific considerations and appropriate strategies (e.g., adjusting environmental conditions, using bioaugmentation, optimizing nutrient supply) can be employed to maximize the effectiveness of bioremediation efforts.

Chapter - 2

Process of Biodegradation and Bioremediation

2.1. Introduction

Biodegradation and bioremediation are essential processes that play a crucial role in the natural breakdown and detoxification of various pollutants in the environment. These processes utilize microorganisms, such as bacteria, fungi, and other biological agents, to break down and transform complex organic compounds into simpler, less harmful substances. Biodegradation and bioremediation are widely employed in addressing environmental pollution and restoring contaminated ecosystems. The continuous release of pollutants, including industrial chemicals, petroleum products, pesticides, plastics, and other hazardous substances, poses a significant threat to the environment and human health. Biodegradation and bioremediation offer sustainable and eco-friendly solutions for managing these pollutants.

Biodegradation refers to the naturally occurring process where microorganisms utilize pollutants as a source of energy and nutrients. These microorganisms break down complex molecules by secreting enzymes that target specific chemical bonds, resulting in the conversion of pollutants into simpler compounds. This process occurs in various environments, including soil, water, and sediments, and is influenced by factors such as temperature, pH, oxygen availability, and the presence of suitable microorganisms. Bioremediation, on the other hand, involves the deliberate manipulation and enhancement of biodegradation processes to accelerate the removal of pollutants from contaminated sites.

Biodegradation and bioremediation offer a sustainable and often cost-effective approach to managing environmental pollution. However, the success of these processes depends on various factors, including the type of pollutant, the environmental conditions, the availability of suitable microorganisms, and the overall strategy

DOI: 10.1201/9781003787549-2

employed. When implemented effectively, these processes can contribute significantly to restoring ecosystems and reducing the impact of pollution on both the environment and human health.

To remediate polluted sites, biological processes have many advantages from economic, environmental, and practical aspects. Adsorption and biodegradation of organic contaminants and the immobilization, mobilization, and/or transformation of metal(loid)s are the main remediation processes that can be mediated by the action of several microorganisms especially those extremophiles surviving in hostile environments with high concentrations of pollutants. Biodegradation or biological degradation is the phenomenon of biological transformation of organic compounds by living organisms, particularly the microorganisms. Biodegradation basically involves the conversion of complex organic molecules to simpler (and mostly non-toxic) ones. The term biotransformation is used for incomplete biodegradation of organic compounds involving one or a few reactions. Biotransformation is employed for the synthesis of commercially important products by microorganisms. The microbes serve as scavengers in bioremediation. The removal of organic wastes by microbes for environmental clean-up is the essence of bioremediation. The other names used (by some authors) for bioremediation are bio-treatment, bio-reclamation and bio-restoration. It is rather difficult to show any distinction between biodegradation and bioremediation. Further, in biotechnology, most of the reactions of biodegradation/ bioremediation involve xenobiotic. For understanding the process of biodegradation requires an understanding of the microorganisms that make the process work. The microbial organisms transform the substance through metabolic or enzymatic processes. It is based on two processes: growth and co-metabolism. In growth, an organic pollutant is used as sole source of carbon and energy. This process results in a complete degradation (mineralization) of organic pollutants. Co-metabolism is defined as the metabolism of an organic compound in the presence of a growth substrate that is used as the primary carbon and energy source. Several microorganisms, including fungi, bacteria and yeasts are involved in biodegradation process. Algae and protozoa reports are scanty regarding their involvement in biodegradation. Biodegradation processes vary greatly, but frequently the final product of the degradation is carbon dioxide. Organic material can be degraded aerobically, with oxygen, or anaerobically, without oxygen.

Biodegradable matter is generally organic material such as plant and animal matter and other substances originating from living organisms, or artificial materials that are similar enough to plant and animal matter to be put to use by microorganisms. Some microorganisms have the astonishing, naturally occurring, microbial catabolic diversity to degrade, transform or accumulate a huge range of compounds including hydrocarbons (e.g. oil), polychlorinated biphenyls (PCBs), polyaromatic hydrocarbons (PAHs), radionuclides and metals.

Bioremediation process can be divided into three phases or levels. First, through natural attenuation, contaminants are reduced by native microorganisms without any human augmentation (Verma and Kuila, 2019). Second, biostimulation is employed where nutrients and oxygen are applied to the systems to improve their effectiveness and to accelerate biodegradation. Finally, during bioaugmentation, microorganisms are added to the systems. These supplemental organisms should be more efficient than native flora to degrade the target contaminant. A feasible remedial technology requires microorganisms being capable of quick adaptation and efficient uses of pollutants of interest in a particular case in a reasonable period of time. Many factors influence microorganisms to use pollutants as substrates or co-metabolize them, like, the genetic potential and certain environmental factors such as temperature, pH, and available nitrogen and phosphorus sources, then, seem to determine the rate and the extent of degradation.

Biodegradation Vs Bioremediation	
Biodegradation is the process of decomposing organic materials in the environment by microorganisms.	Bioremediation is a waste management technique that uses biological agents to clean the contaminants in the environment.
It is a natural process that happens without human intervention.	It is an engineered process that happens with human intervention is used to control the rate of bioremediation, by control of temperature, availability of food or nutrients, etc.
This is a slow process.	This is a faster process.
It is both beneficial and harmful.	It always has beneficial effects.
It happens everywhere in the environment.	Bioremediation is planned at a contaminated site.
There is no need for experts.	Experts are required to design and implement this process.

These processes play a crucial role in cleaning up polluted sites and restoring ecosystems. Here's an overview of the processes of biodegradation and bioremediation:

2.2. Steps of Biodegradation

The biodegradation process involves several steps:

i. Enzymatic Hydrolysis

Enzymatic hydrolysis is a process that involves the use of enzymes to break down complex molecules into simpler components through the addition of water molecules. In the context of biodegradation, enzymatic hydrolysis plays a crucial role in breaking down complex organic compounds into smaller, more readily biodegradable molecules.

Enzymatic hydrolysis enhances the rate of biodegradation by facilitating the breakdown of complex molecules that may be resistant to microbial action. Enzymes are specialized proteins that act as biological catalysts, speeding up chemical reactions without being consumed in the process. In the context of biodegradation, enzymes target specific chemical bonds within complex organic molecules and break them apart, resulting in the formation of simpler compounds that can be more easily metabolized by microorganisms (Saravanan et al., 2021).

For example, cellulose is a complex carbohydrate found in plant cell walls and other organic materials. It is composed of many glucose molecules linked together through strong chemical bonds. Microorganisms that can break down cellulose, such as certain bacteria and fungi, produce enzymes called cellulases. These cellulases hydrolyze the glycosidic bonds between the glucose molecules, converting cellulose into glucose and other simple sugars that the microorganisms can then use as a source of energy.

Enzymatic hydrolysis is also used in various industrial processes. For instance, in the production of biofuels like ethanol from lignocellulosic biomass, enzymatic hydrolysis is a critical step. The complex carbohydrates present in the biomass are broken down into fermentable sugars through enzymatic hydrolysis, which are then converted into ethanol by fermentation.

Here are the steps involved in enzymatic hydrolysis during biodegradation:

Recognition and Binding: Enzymes are specialized proteins that catalyze specific chemical reactions. Microorganisms release enzymes that can recognize and bind to specific chemical bonds within the complex organic molecule they want to degrade. Enzymes are highly specific, so they only target certain types of bonds.

i. *Formation of Enzyme-Substrate Complex:* When the enzyme binds to the target molecule at its active site (the region where the reaction takes place), an enzyme-substrate complex is formed. This complex brings the reacting molecules in close proximity, creating an environment where the bond to be hydrolyzed is more susceptible to breakage.

ii. *Hydrolysis Reaction:* Hydrolysis is a chemical reaction that involves the cleavage of a chemical bond by the addition of water molecules. In the context of enzymatic hydrolysis during biodegradation, the enzyme facilitates the addition of a water molecule to the bond being targeted. This results in the bond breaking and the formation of two smaller molecules, each containing one of the fragments from the original bond.

iii. *Product Release:* Once the hydrolysis reaction occurs, the enzyme releases the newly formed products. These products are now smaller and more water-soluble than the original complex organic molecule, making them easier for microorganisms to transport and utilize.

iv. *Enzyme Recycling:* After the hydrolysis reaction is complete and the products are released, the enzyme is not consumed in the reaction. Instead, it can continue to catalyze additional hydrolysis reactions on other molecules. Enzymes are highly efficient catalysts and can facilitate many rounds of reactions.

v. *Microbial Uptake and Metabolism:* The products of enzymatic hydrolysis are often simpler compounds that microorganisms can more easily transport across their cell membranes. Once inside the microbial cell, these compounds can be further metabolized to generate energy and synthesize cellular components.

ii. Metabolic Reactions

The metabolic reactions during biodegradation can be broadly categorized into three main stages: extracellular enzyme action, intracellular metabolism, and waste product formation (Moreno and Rojo, 2023).

i. *Extracellular Enzyme Action*: The process of biodegradation often begins with the secretion of extracellular enzymes by microorganisms. These enzymes target specific bonds within complex organic molecules, such as carbohydrates, proteins, and lipids. For example, lipases break down fats and oils, proteases break down proteins, and amylases break down starches. These extracellular enzymes break down larger molecules into smaller components that can be transported into the microbial cells for further metabolism.

ii. *Intracellular Metabolism*: Once the smaller molecules enter the microbial cells, intracellular metabolism takes place. The breakdown products are further processed through a series of metabolic pathways, including glycolysis, the citric acid cycle (also known as the Krebs cycle), and oxidative phosphorylation. These pathways involve a series of enzymatic reactions that gradually extract energy from the broken-down molecules. The energy generated is captured in the form of adenosine triphosphate (ATP), a molecule that provides energy for cellular processes.

Different microorganisms use different pathways and enzymes depending on the types of compounds they encounter. For instance, carbohydrates are typically metabolized through glycolysis, proteins through proteolysis, and fats through β-oxidation.

iii. *Waste Product Formation*: As the intracellular metabolism progresses, waste products are generated. These waste products include carbon dioxide, water, and other simpler compounds. Carbon dioxide is a common end product of the breakdown of organic molecules, and it is released into the environment. Water is also produced as a result of various metabolic reactions. These waste products are either released into the surrounding environment or used by the microorganisms for their own growth and maintenance.

iii. Mineralization

Mineralization is a crucial process in the biodegradation of organic materials, particularly in the context of environmental remediation and waste management (Qin et al., 2020). Biodegradation refers to the breakdown of organic substances by microorganisms into simpler compounds like carbon dioxide, water, and inorganic minerals. Mineralization is the final step in this process, where the remaining organic molecules are fully converted into inorganic forms, primarily carbon dioxide, water, and various minerals or salts.

The mineralization process involves the complete transformation of organic compounds, which may include complex molecules like proteins, carbohydrates, and lipids, into simpler inorganic compounds. This process is mainly carried out by microorganisms such as bacteria, fungi, and other decomposers that feed on the organic material.

Steps involved in mineralization during biodegradation

i. *Initial Breakdown*: In the biodegradation process, microorganisms first break down complex organic molecules into simpler compounds through enzymatic reactions. This often results in the formation of smaller organic molecules, such as organic acids and alcohols.

ii. *Secondary Metabolism*: Microorganisms then further metabolize these intermediate products. During this stage, many of the remaining organic compounds are oxidized, releasing energy for the microorganisms' growth and survival.

iii. *Final Conversion*: As the microorganisms continue to metabolize the intermediate products, they eventually convert the remaining organic compounds into their simplest forms: carbon dioxide (CO_2), water (H_2O), and inorganic minerals or salts. This is the mineralization stage.

iv. *Release of Inorganic Products*: The inorganic products, such as CO_2 and water, are released into the environment. The minerals or salts formed as a result of mineralization can become part of the soil or water composition.

It's important to note that the rate of mineralization can vary depending on factors such as the type of organic material, the composition of the microbial community, environmental conditions (temperature, moisture, pH), and the availability of nutrients. In some cases, certain compounds might resist complete mineralization due to their chemical structure, and these residues could persist in the environment for longer periods.

Bioremediation involves the introduction of specific microorganisms or the enhancement of existing microbial communities to accelerate the degradation of pollutants.

Step by step process of bioremediation (Ayilara and Babalola, 2023)

i. *Site Characterization and Assessment:* The first stage involves a thorough investigation of the contaminated site. This includes identifying the nature and extent of the contamination, determining the types of pollutants present, and assessing the potential risks to the environment and human health. Site characterization helps in understanding the conditions and factors that may influence the bioremediation process.

ii. *Feasibility Study and Planning:* Based on the site assessment, a feasibility study is conducted to determine if bioremediation is a suitable and effective remediation option for the specific site and contamination. The study evaluates factors such as the type of pollutants, their concentrations, the site's hydrogeology, and the available resources. If bioremediation is deemed feasible, a detailed remediation plan is developed, outlining the specific bioremediation approach, the choice of microorganisms or plants, and the timeline for implementation.

iii. *Preparation and Site Setup:* Before initiating bioremediation, the site needs to be prepared appropriately. This may involve clearing the area of debris and ensuring safety measures are in place for the workers and the surrounding environment. For ex situ bioremediation, the contaminated soil or water may need to be excavated and transported to a treatment facility.

iv. Bioremediation Treatment: The actual bioremediation process is the core stage of the entire remediation project. Depending on the type of contaminants and the site conditions, there are two primary methods of bioremediation:

(a) In Situ Bioremediation: In this method, microorganisms or plants are introduced directly into the contaminated soil or water at the site. The selected organisms work to degrade, metabolize, or absorb the pollutants, converting them into less harmful substances. (*The detailed has been described in Chapter No.3)*

(b) Ex Situ Bioremediation: In this approach, the contaminated material is excavated and transported to a controlled environment, such as a bioreactor or treatment facility. The bioremediation process occurs in this controlled setting, and the treated material may be returned to the site or disposed of responsibly. (*The detailed has been described in Chapter No. 4)*

During the bioremediation treatment, environmental factors such as temperature, moisture, pH levels, oxygen availability, and nutrient concentrations are closely monitored and controlled to optimize the activity of the chosen microorganisms or plants.

» **Monitoring and Adjustment**

Throughout the bioremediation process, regular monitoring is conducted to evaluate the progress of contaminant degradation or removal. Samples are collected and analyzed to assess the effectiveness of the bioremediation treatment and make any necessary adjustments to the strategy if required.

» **Completion and Verification**

Once the desired cleanup objectives are achieved, and contaminant levels have been reduced to acceptable levels, the bioremediation process is considered complete. Verification of the successful cleanup is conducted through further site assessment and sampling to ensure that the site meets the predetermined cleanup criteria and regulatory standards.

» **Post-Remediation Monitoring and Maintenance**

After completing the bioremediation process, post-remediation monitoring may continue for a certain period to ensure the long-term effectiveness of the treatment. Additionally, maintenance measures may be implemented to support the natural recovery and stabilization of the site.

By following these step-by-step stages, bioremediation can be an effective and environmentally friendly approach to clean up contaminated sites.

Mechanism of Bioremediation

Bioremediation is a natural or engineered process that uses living organisms (usually microorganisms) to degrade, remove, or neutralize contaminants in the environment (Tegene, 2020). The process can be applied to various types of pollutants, including organic compounds, heavy metals, and other toxic substances. Here's a detailed mechanism of bioremediation:

i. Identification and Selection of Microorganisms: The first step in bioremediation is to identify and select microorganisms that are capable of degrading the specific contaminants present in the contaminated site. These microorganisms can be naturally occurring in the environment or may be isolated from other contaminated sites with similar pollutants.

ii. Nutrient Addition: To encourage microbial growth and activity, nutrients like nitrogen, phosphorus, and other essential elements are often added to the contaminated site. These nutrients act as stimulants for the growth of the selected microorganisms, enhancing their ability to degrade pollutants.

iii. Contaminant Uptake: The selected microorganisms utilize the contaminants as a source of energy or nutrients for their growth. They take up the pollutants through various mechanisms, such as passive diffusion or active transport, depending on the nature of the contaminants.

iv. Biodegradation: Once the contaminants are taken up by the microorganisms, enzymatic processes inside the cells break down the complex compounds into simpler molecules. These enzymes act as catalysts, accelerating the biodegradation process. The end products of this biodegradation are usually less harmful compounds, such as carbon dioxide, water, and non-toxic substances.

v. Environmental Factors: Several environmental factors play a crucial role in the success of bioremediation. Temperature, pH, oxygen availability, moisture content, and other factors must be carefully monitored and controlled to create an optimal environment for microbial activity.

vi. Bioaugmentation (Optional): In some cases, the natural population of microorganisms might not be sufficient to effectively remediate the contamination. In such situations, bioaugmentation is employed, where additional beneficial microorganisms are introduced to the contaminated site to boost biodegradation.

vii. Biostimulation (Optional): If the existing microbial population is present but not sufficiently active, biostimulation is used to enhance their activity. This can involve adding specific compounds or substances that stimulate the growth and metabolic activity of the native microorganisms.

viii. Monitoring and Optimization: Throughout the bioremediation process, regular monitoring of contaminant levels and microbial activity is essential. Adjustments might be made based on the data collected to optimize the bioremediation process.

ix. End-Product Monitoring: After bioremediation is complete or significantly reduced, monitoring continues to ensure that the contaminants have been successfully degraded or removed to acceptable levels, and no harmful by-products have accumulated.

x. Long-Term Monitoring: After the completion of bioremediation, long-term monitoring is often conducted to ensure that the site remains free of contaminants and that there are no signs of recontamination.

2.5. Acclimation

Acclimation is the process by which an organism adjusts its physiological, behavioral, or biochemical traits in response to changes in its environment. It allows the organism to cope with new or altered environmental conditions, helping it to survive and function optimally (Wang et al., 2018). Acclimation is a reversible and relatively rapid adaptation to a changed environment. The primary purpose of acclimation is to maintain the internal balance or homeostasis of an organism in response to external changes. By adjusting their physiological processes, organisms can optimize their performance, reproduction, and overall survival in the new environment. The lag phase is the initial phase of bacterial growth, during which bacteria are adjusting to their new environment, synthesizing essential enzymes, and preparing for rapid growth. Acclimatization primarily occurs during this phase because bacteria need time to respond to and adapt to the new environmental conditions before they can begin to multiply rapidly.

Microbes are incredibly adaptable and versatile organisms. They exhibit various mechanisms of acclimation to respond to changes in their environment. Acclimation in microbes allows them to survive and thrive in diverse and challenging conditions, ranging from extreme temperatures and pH levels to varying nutrient availability.

Acclimation is different from adaptation, as acclimation is a reversible process that occurs within the lifetime of an individual, whereas adaptation involves genetic changes that occur over generations. Adaptation and acclimation are both biological processes that allow organisms to adjust to changes in their environment. However, they differ in their time frame, mechanism, and scope of changes. Here are the key differences between adaptation and acclimation:

2.5.1. Time Frame:

- *Adaptation*: Adaptation occurs over multiple generations through genetic changes. It involves the selection and inheritance of advantageous traits that become more prevalent in a population over time. This process is relatively slow and occurs over evolutionary timescales.
- *Acclimation*: Acclimation takes place within the lifetime of an individual organism. It involves reversible physiological, behavioral, or biochemical adjustments that occur in response to changes in the immediate environment. Acclimation is a rapid process and enables the organism to cope with short-term environmental variations.

2.5.2. Genetic Basis:

- *Adaptation*: Genetic changes are the basis of adaptation. Mutations and natural selection drive the evolution of advantageous traits that provide better fitness and survival advantages to individuals in a population.
- *Acclimation*: Acclimation is a non-genetic process and does not involve changes in the organism's DNA. Instead, it relies on pre-existing genetic diversity and the organism's ability to modulate its physiological responses to the environment.

2.5.3. Reversibility:

- *Adaptation*: Adaptations are generally irreversible within an individual's lifetime. Once a genetic trait is fixed in a population, it will persist in subsequent generations.
- *Acclimation*: Acclimation is reversible. If the environmental conditions change back to their original state, the organism can readjust its traits accordingly.

2.5.4. Scope of Changes:

- *Adaptation*: Adaptation results in permanent changes in the population's gene pool and can lead to the emergence of entirely new traits that enhance survival and reproduction.
- *Acclimation*: Acclimation involves temporary adjustments that help an individual cope with immediate environmental fluctuations. These changes are specific to the individual and do not alter the genetic makeup of the population.

2.5.5. Time Scale:

- *Adaptation*: Adaptations occur over long periods, often requiring many generations for significant changes to become widespread in a population.
- *Acclimation*: Acclimation occurs over short periods, within the lifetime of an individual, and is usually related to changes in the individual's environment.

2.5.6. Types of acclimation demonstrated by microbes:

i. Temperature Acclimation:

Temperature acclimation by microbes refers to the process by which these microorganisms adjust their physiological and molecular responses to cope with changes in temperature in their environment. Microbes are highly adaptable to different temperature ranges, and temperature acclimation plays a crucial role in their survival and growth in diverse habitats. Here are some key aspects of temperature acclimation by microbes:

- *Membrane Composition:* Microbes modify the composition of their cell membranes in response to temperature changes. As temperatures decrease, they may increase the proportion of unsaturated fatty acids in their membranes to maintain membrane fluidity. Conversely, at higher temperatures, they might incorporate more saturated fatty acids to enhance membrane stability.

- *Enzyme Production and Activity:* Microbes adjust the production and activity of enzymes based on temperature variations. Cold-acclimated microbes produce enzymes with higher catalytic efficiency at lower temperatures (psychrophilic enzymes), while heat-acclimated microbes produce thermophilic enzymes that remain stable and functional at high temperatures.
- *Heat Shock Proteins:* When exposed to sudden temperature increases, microbes synthesize heat shock proteins (HSPs) to protect cellular structures and proteins from heat-induced denaturation. These proteins assist in refolding damaged proteins and maintaining cellular integrity under stress.
- *Cold Shock Proteins:* In response to sudden temperature decreases, microbes produce cold shock proteins to help stabilize RNA molecules and prevent the formation of secondary structures that could hinder translation.
- *Metabolic Adjustments:* Temperature acclimation can lead to changes in the metabolic pathways of microbes. For example, at lower temperatures, some microbes may preferentially utilize specific carbon sources or alter their energy generation processes to adapt to the reduced metabolic rates.
- *DNA Repair and Replication:* Microbes acclimate to different temperatures by adjusting their DNA repair and replication processes. This ensures accurate DNA replication and repair of heat-induced DNA damage.
- *Chaperones and Proteases:* Microbes produce molecular chaperones and proteases that assist in the proper folding and assembly of proteins, especially when the temperature changes stress protein folding.
- *Cell Size and Surface Area-to-Volume Ratio:* Some microbes alter their cell size and shape in response to temperature changes to optimize their surface area-to-volume ratio, which influences their heat exchange and nutrient uptake capabilities.

ii. pH Acclimation:

pH acclimation by microbes refers to the ability of these microorganisms to adjust their internal pH and physiological responses in response to changes in the acidity or alkalinity of their environment. pH is a critical environmental factor that affects the chemical properties of the surrounding medium and can influence microbial growth, metabolism, and survival. Microbes exhibit various strategies to acclimate to different pH levels, enabling them to thrive in a wide range of habitats. Here are some key aspects of pH acclimation by microbes:

» *pH Homeostasis*: Microbes maintain their internal pH within a narrow range, a process known as pH homeostasis. They achieve this by using proton pumps, ion transporters, and other pH-regulating mechanisms to expel or take up protons ($H+$) as needed, depending on the external pH conditions.

» *Ion Exchange*: Microbes can exchange ions across their cell membranes to regulate internal pH. For example, they may take up potassium ions ($K+$) in exchange for protons to neutralize the acidity of the cytoplasm.

» *Acid Tolerance Response* (ATR): Some microbes possess an acid tolerance response, which involves the induction of specific genes and the production of protective proteins, such as chaperones and enzymes, to counteract the detrimental effects of acidic conditions on cellular components.

» *Alkaline Tolerance Response*: Similarly, microbes exposed to alkaline conditions may activate an alkaline tolerance response, involving the upregulation of genes associated with pH homeostasis and protection against alkali-induced stress.

» *Extracellular Buffering*: Some microbes produce extracellular buffering compounds, such as organic acids or ammonia, to modulate the pH of their surroundings and create a more favorable environment for growth.

» *pH-Responsive Proteins*: Microbes produce pH-responsive proteins that undergo conformational changes in response to pH fluctuations. These proteins may function optimally at specific pH levels, ensuring the maintenance of essential cellular processes under varying pH conditions.

» *Acid Resistance Mechanisms*: Acidophiles, microbes that thrive in extremely acidic environments, have specialized mechanisms to resist the harmful effects of low pH. They may possess acid-resistant enzymes and proteins, as well as cell membrane modifications to withstand the acidic conditions.

» *Alkaline Resistance Mechanisms*: Alkaliphiles, microbes adapted to alkaline environments, employ specific alkaline resistance mechanisms to cope with the challenges posed by high pH levels. These mechanisms help protect cellular structures and maintain functional integrity.

iii. Oxygen Acclimation:

Oxygen acclimation by microbes refers to the process by which microorganisms adjust their physiological and metabolic responses to changes in the availability of oxygen in their environment. Oxygen is a critical environmental factor that influences microbial growth, metabolism, and survival. Some microbes require oxygen for respiration, while others are adapted to thrive in oxygen-depleted or oxygen-free environments. Here are some key aspects of oxygen acclimation by microbes:

» *Aerobic and Anaerobic Respiration*: Microbes that rely on oxygen for respiration are known as aerobic organisms. They possess enzymes and metabolic pathways that facilitate the use of oxygen as the final electron acceptor in the electron transport chain, allowing for efficient energy generation through oxidative phosphorylation. In contrast, anaerobic microbes can acclimate to oxygen-depleted environments and use alternative electron acceptors, such as nitrate, sulfate, or carbon dioxide, for anaerobic respiration.

» *Oxygen Sensing and Signal Transduction*: Microbes have oxygen-sensing systems that allow them to detect changes in oxygen levels in their surroundings. They can then initiate specific cellular responses through signal transduction pathways to acclimate to the prevailing oxygen conditions.

» *Aerobic and Anaerobic Enzymes*: Microbes adjust their enzyme expression patterns based on oxygen availability. Aerobic organisms upregulate enzymes involved in oxidative metabolism, such as cytochrome oxidases and superoxide dismutases, to protect against reactive oxygen species (ROS) generated during respiration. Anaerobic organisms, on the other hand, produce specific enzymes for anaerobic respiration or fermentation.

» *Oxygen Radicals and ROS Detoxification*: Oxygen can be both beneficial and harmful to microbes. Aerobic respiration generates ROS, which can damage cellular components. Microbes employ antioxidant enzymes, such as catalase and peroxidase, to neutralize ROS and protect themselves from oxidative stress.

» *Oxygen Diffusion and Microbial Aggregates*: Oxygen concentration can vary significantly within microbial communities. Microbes within aggregates or biofilms may experience different oxygen levels depending on their position within the structure, leading to differential acclimation responses.

» *Microaerophiles*: Some microbes are classified as microaerophiles, which require low levels of oxygen for optimal growth. They acclimate to environments with limited oxygen availability and possess specific mechanisms to protect themselves from excessive oxygen exposure.

» *Obligate Anaerobes*: Obligate anaerobes are microbes that cannot tolerate oxygen and may even be harmed or killed by its presence. They are adapted to live in oxygen-free environments and rely on anaerobic metabolism for energy production.

iv. Nutrient acclimation

Nutrient acclimation by microbes refers to the process by which microorganisms adjust their physiological and molecular responses to cope with changes in the availability of nutrients in their environment. Nutrients are essential for microbial growth, metabolism, and survival, and their availability can fluctuate in various habitats. Microbes exhibit various strategies to acclimate to different nutrient conditions, enabling them to thrive in a wide range of ecological niches. Here are some key aspects of nutrient acclimation by microbes:

» *Metabolic Plasticity*: Microbes possess metabolic plasticity, allowing them to adjust their metabolic pathways and nutrient utilization strategies based on nutrient availability. For example, they may switch between different carbon sources or nitrogen sources depending on what is readily available in the environment.

» *Nutrient Uptake Mechanisms*: Microbes may alter their nutrient uptake mechanisms to efficiently acquire available nutrients. They may increase the expression of nutrient transporter proteins on their cell membranes to enhance nutrient uptake rates.

» *Enzyme Production*: In response to variations in nutrient availability, microbes can adjust the production of specific enzymes required for nutrient metabolism. For example, they may synthesize enzymes that can degrade complex organic molecules into simpler compounds for easier uptake and utilization.

» *Nutrient Storage:* Some microbes have the ability to store excess nutrients as intracellular reserves, such as glycogen for carbon and polyphosphate for phosphorus. These reserves can be utilized when nutrients become limited, providing a survival advantage.

» *Nitrogen Fixation:* Certain microbes have the capacity to fix atmospheric nitrogen into forms usable by other organisms. They acclimate to nitrogen-poor environments by increasing nitrogen fixation activities, which is crucial for enriching the nitrogen content in the ecosystem.

» *Phosphate Acquisition:* Microbes can adjust their phosphate acquisition mechanisms to cope with limited phosphate availability in their surroundings. They may produce phosphatases to scavenge organic phosphorus sources or form symbiotic relationships with other organisms to access phosphate.

» *Biofilm Formation:* In nutrient-limited conditions, some microbes form biofilms-complex communities of microorganisms embedded in a self-produced extracellular matrix. Biofilms enhance nutrient retention and resource sharing among community members, promoting survival in nutrient-poor environments.

» *Quorum Sensing:* Microbes use quorum sensing to sense population density and adjust their behavior and metabolic activities accordingly. In nutrient-depleted environments, quorum sensing may trigger the expression of genes involved in nutrient acquisition or stress response.

v. Osmotic acclimation:

Osmotic acclimation is a process by which microbes adjust their physiological and cellular mechanisms to cope with changes in osmotic pressure in their surrounding environment. Osmotic pressure refers to the pressure exerted by the concentration of solutes (such as salts) in a solution. Microbes encounter various osmotic conditions, from low to high osmotic pressure, and they must adapt to these changes to survive and thrive.

Here's how osmotic acclimation is demonstrated by microbes:

» *Osmoregulation*: Microbes can regulate the concentration of osmolytes (small organic molecules) within their cells to maintain an optimal internal osmotic pressure that balances the external environment. Osmolytes are molecules that help stabilize cellular structures and enzymes under varying osmotic conditions. Examples of osmolytes include proline, glycine betaine, and trehalose.

» *Compatible solute accumulation*: In response to high osmotic stress, some microbes accumulate compatible solutes, which are osmolytes that do not disrupt cellular processes even at high concentrations. These compatible solutes help protect cellular structures from dehydration and stabilize cellular components under osmotic stress.

» *Membrane adjustments*: The cell membrane is a crucial barrier that separates the internal cellular environment from the external surroundings. Microbes can modify the composition and fluidity of their cell membranes to adapt to different osmotic conditions. In high osmotic conditions, they may synthesize more lipids or alter their lipid composition to reduce water loss and maintain membrane integrity.

- *Aquaporins*: Aquaporins are specialized proteins that facilitate the movement of water across cell membranes. Some microbes can regulate the expression of aquaporins to control water flow and maintain cellular hydration.
- *Osmosensing and signal transduction*: Microbes possess osmosensing mechanisms that enable them to detect changes in osmotic conditions. When changes are detected, signal transduction pathways are activated to trigger appropriate responses, such as osmolyte synthesis or adjustment of membrane permeability.
- *Halotolerance*: Halotolerant microbes can withstand high salt concentrations without being negatively affected. They achieve this through various strategies, including the accumulation of compatible solutes, adjustments in membrane composition, and efficient ion transport systems.

vi. Desiccation acclimation

Desiccation acclimation, also known as desiccation tolerance or desiccation resistance, is a phenomenon exhibited by certain microbes, particularly extremophiles, that allows them to survive periods of extreme dryness or dehydration. Desiccation is the process of drying out, and it can be a significant environmental challenge for microorganisms that encounter it in various habitats, such as arid regions, desert soils, or even on surfaces like rocks or building materials. Here's how microbes demonstrate desiccation acclimation:

- *Osmotic adjustment*: During desiccation, water is lost from microbial cells, leading to cellular dehydration. To counteract this, desiccation-acclimated microbes can adjust their internal osmotic balance by accumulating osmolytes (compatible solutes) to help retain cellular water and protect essential cellular structures.
- *Synthesis of protective molecules*: Microbes undergoing desiccation acclimation may synthesize specific protective molecules, such as trehalose and certain proteins (e.g., late embryogenesis abundant proteins, or LEAPs). These molecules help stabilize cellular components and enzymes, preventing damage during dehydration and rehydration.

- *Formation of biofilms and cysts*: Some desiccation-acclimated microbes form biofilms or cysts during periods of drought or desiccation. Biofilms are communities of microorganisms embedded in a protective extracellular matrix, while cysts are dormant, tough-walled structures that can withstand harsh conditions until more favorable conditions return.
- *Antioxidant defense*: During desiccation, reactive oxygen species (ROS) can accumulate, causing cellular damage. Desiccation-acclimated microbes often produce high levels of antioxidants to neutralize ROS and protect their cellular components.
- *DNA repair and protection*: Desiccation can lead to DNA damage, so desiccation-acclimated microbes have evolved efficient DNA repair mechanisms and protective proteins to safeguard their genetic material.
- *Reduced metabolic activity*: In response to desiccation, many microbes can enter a dormant or quiescent state, reducing their metabolic activity to conserve energy and minimize water loss.
- *Cellular water retention*: Some desiccation-acclimated microbes have unique cell membrane structures that aid in water retention and prevent excessive water loss during dehydration.
- *Spore formation*: Certain microbes, like bacterial endospores, can undergo sporulation to create highly resistant spores with tough protective coats. These spores can survive desiccation and harsh environmental conditions until more favorable conditions arise.

vii. Radiation acclimation

Radiation acclimation, also known as radioresistance or radiation tolerance, is a remarkable adaptation demonstrated by certain microbes that enables them to withstand and survive exposure to ionizing radiation. Ionizing radiation, such as gamma rays, X-rays, and high-energy particles, can cause severe damage to cellular structures and DNA. Microbes with radiation acclimation have developed various mechanisms to cope with and repair the damage caused by radiation. Here are some ways microbes demonstrate radiation acclimation:

- *DNA repair mechanisms*: Radiation can induce single-strand breaks, double-strand breaks, and other types of DNA damage. Radiation-acclimated microbes possess efficient DNA repair mechanisms, including base excision repair, nucleotide excision repair, homologous recombination, and nonhomologous end joining, which help restore damaged DNA and maintain genomic stability.

- *Radical scavenging*: Ionizing radiation generates reactive oxygen species (ROS), such as free radicals, which can cause additional damage to cellular components. Radiation-acclimated microbes produce high levels of antioxidant enzymes and molecules that scavenge ROS, reducing oxidative stress and minimizing cellular damage.
- *Protein protection and repair*: Radiation can lead to protein denaturation and damage. Some radiation-acclimated microbes produce specialized chaperones and proteases that aid in protein refolding and repair.
- *DNA damage tolerance*: In addition to repair mechanisms, radiation-acclimated microbes may possess mechanisms to tolerate some level of DNA damage, allowing them to function despite the presence of lesions in their genetic material.
- *Radical resistance*: Some radiation-acclimated microbes have specific biochemical pathways that resist the damaging effects of radicals formed during radiation exposure.
- *Metal ion uptake*: Certain microbes use metal ions, such as manganese, as cofactors for enzymes involved in radiation protection and DNA repair. The presence of these ions enhances their ability to withstand radiation.
- *Efficient cell cycle control*: Radiation-acclimated microbes tightly regulate their cell cycle, enabling them to halt cell division when DNA damage occurs and initiate repair mechanisms before continuing replication.
- *Adaptive response*: Some microbes exhibit an adaptive response to low doses of radiation, which induces an increased resistance to subsequent higher doses. This phenomenon helps prepare the cells for potential future exposure to higher radiation levels.

viii. Pressure acclimation

Pressure acclimation, also known as barotolerance or piezotolerance, is the ability of certain microbes to adapt and thrive in environments with high pressure conditions, such as the deep sea or deep subsurface environments. These microbes have evolved specific mechanisms to withstand the extreme hydrostatic pressures found at great depths. Here are some ways microbes demonstrate pressure acclimation:

» *Membrane adaptations*: High pressure can affect the fluidity and stability of cellular membranes. Pressure-acclimated microbes may adjust the composition of their cell membranes to maintain integrity and prevent leakage of cellular contents.

» *Enzyme stability*: Pressure can denature proteins and enzymes, affecting cellular metabolism. Pressure-acclimated microbes produce enzymes with enhanced stability, enabling them to function effectively under high-pressure conditions.

» *Baroprotectants*: Some microbes synthesize special compounds called baroprotectants, which help protect cellular structures from the detrimental effects of pressure. These compounds may act as osmolytes or stabilize cellular macromolecules.

» *DNA repair mechanisms*: High pressure can cause damage to DNA, leading to mutations and cell death. Pressure-acclimated microbes have efficient DNA repair systems to correct such damage.

» *Regulation of gene expression*: Microbes can adjust their gene expression profiles in response to pressure changes. Certain genes related to pressure acclimation are upregulated, enabling the synthesis of proteins and protective molecules required for survival under high pressure.

» *Piezophiles*: Some microbes, known as piezophiles or barophiles, are specifically adapted to high-pressure environments. They thrive under extreme pressure conditions and may not survive at normal atmospheric pressure.

» *Structural adaptations*: Pressure-acclimated microbes may exhibit changes in their cell wall structure, flagella, or other surface appendages, which are essential for their survival and motility in high-pressure environments.

» *Chaperones and proteases*: Pressure-acclimated microbes produce specific chaperone proteins and proteases that assist in the folding and repair of damaged proteins caused by pressure-induced stress.

ix. Toxic substance acclimation

Toxic substance acclimation, also known as resistance or tolerance, is the ability of certain microbes to adapt and survive in the presence of toxic substances that would be lethal or inhibitory to other organisms (Di et al., 2020). Microbes encounter a wide range of toxic compounds in their environments, such as heavy metals, organic pollutants, antibiotics, and other chemicals. Here are some ways microbes demonstrate toxic substance acclimation:

- *Efflux pumps*: Microbes can express specialized membrane transport proteins called efflux pumps, which actively pump toxic substances out of their cells. This helps maintain low intracellular concentrations of toxic compounds and prevents their accumulation to lethal levels.
- *Enzymatic detoxification*: Some microbes produce specific enzymes that break down or transform toxic substances into less harmful forms. For example, certain bacteria possess metal detoxification systems that convert toxic heavy metals into less toxic complexes.
- *Biochemical resistance mechanisms*: Microbes may possess unique biochemical pathways that neutralize or sequester toxic compounds, rendering them harmless to the cell.
- *Genetic mutations*: In some cases, microbes may acquire genetic mutations that confer resistance to toxic substances. These mutations can alter the structure or function of cellular targets, reducing the binding affinity of toxic compounds.
- *Biofilm formation*: Microbes can form biofilms, which are communities of cells encased in a protective extracellular matrix. Biofilms provide physical protection and can limit the diffusion of toxic substances, increasing the overall resistance of the microbial community.
- *Horizontal gene transfer*: Microbes can exchange genetic material through processes like conjugation, transduction, or transformation. This allows them to acquire resistance genes from other organisms, rapidly spreading resistance traits within a population.
- *Adaptation through exposure*: Repeated exposure to sublethal levels of toxic substances can lead to acclimation, where microbes gradually become more resistant to the toxic compounds over time.
- *Cellular compartmentalization*: Microbes may sequester toxic substances within specific cellular compartments or vesicles, limiting their interactions with vital cellular components.
- *Metabolic bypass*: Some microbes can develop alternative metabolic pathways that allow them to avoid the toxic effects of certain substances or use them as a source of energy.

2.6. Bacterial Growth and Growth Kinetics

Microbial growth is defined as an increase in the number of bacterial cells or an increase in the total mass (Jannasch and Egli, 1993). Bacterial growth is a fundamental biological process wherein bacterial cells multiply and increase in number over time. Understanding the growth of bacteria is crucial in various fields, including microbiology, medicine, food industry, environmental science, and biotechnology. Bacterial growth is a complex process involving numerous anabolic (synthesis of cell constituents and metabolites) and catabolic (breakdown of cell constituents and metabolites) reactions.

The cells increase in population through the mode of asexual reproduction, and since no event occurs (such as recombination and crossing between two parents), the produced cells through this mode are identical to each other. Bacterial growth is influenced by various factors, including temperature, pH, nutrient availability, oxygen levels, and the presence of inhibitory substances. Understanding these factors helps in designing optimal conditions for promoting or inhibiting microbial growth, depending on the specific application.

2.6.1. Type of Reproduction in Bacteria

Bacteria primarily reproduce through asexual reproduction, and the most common form of asexual reproduction in bacteria is binary fission (Bradley, 2019). However, some processes in bacteria, including bacterial conjugation, transformation and transduction, may be considered analogous to sexual reproduction in that they incorporate new genetic information.

Let's explore the different types of bacterial reproduction:

i. Asexual reproduction

Binary fission

Binary fission is the primary mode of asexual reproduction in bacteria. It is a simple and rapid process that allows bacteria to multiply quickly and efficiently. Here's a detailed explanation of binary fission in bacteria:

Initiation of Reproduction: Binary fission starts with a single bacterial cell, often referred to as the "parent cell." The parent cell contains a single circular chromosome (in most bacteria), which carries the genetic material necessary for the survival and functioning of the bacterium.

DNA Replication: Before cell division occurs, the bacterial chromosome undergoes replication. The replication process produces two identical copies of the chromosome, each containing the same genetic information. The two replicated DNA molecules remain attached to the cell membrane.

Cell Elongation: As the cell prepares for division, it elongates in length. This elongation is necessary to accommodate the two copies of the replicated DNA, which will later be segregated into the daughter cells.

Formation of Septum: Next, a septum (cell wall) starts to form at the midpoint of the cell, dividing it into two compartments. The septum is made up of new cell wall material synthesized by the bacterium.

Segregation of DNA: As the septum continues to grow inward, it effectively separates the two copies of the replicated DNA, ensuring that each daughter cell will receive one copy of the genetic material.

Completion of Division: Eventually, the septum pinches off the two compartments, fully dividing the parent cell into two identical daughter cells. Each daughter cell now contains one copy of the replicated genetic material.

Cell Separation: Once the two daughter cells have fully separated, they are released into the environment or remain in close proximity, depending on the bacterial species and the surrounding conditions.

Independent Growth: Each daughter cell now has the potential to grow and divide independently. They will enter the next round of the bacterial cell cycle, where they will continue to elongate, replicate their DNA, and undergo binary fission, leading to exponential population growth.

It is important to note that binary fission is a clonal process, meaning that the offspring (daughter cells) are genetically identical to the parent cell. As a result, bacterial populations can rapidly increase in number under favorable environmental conditions, contributing to their high adaptability and widespread distribution in various habitats. The time required for one complete cycle of binary fission, known as the generation time, can vary depending on the bacterial species and the surrounding environmental conditions. Some bacteria can complete a generation cycle in as little as 20 minutes, while others may take several hours or even days.

Budding

Budding is a form of asexual reproduction commonly observed in certain types of bacteria and other organisms (Sahni and Pei, 2024). During budding, a new organism or offspring develops as an outgrowth or "bud" from the parent organism. This process results in the creation of genetically identical offspring, as the genetic material is copied and passed on without involving the fusion of gametes (sex cells).

In bacterial budding, the following steps typically occur:

Initiation: The parent bacterial cell starts to grow and replicates its genetic material (DNA) in preparation for division.

Budding: A small protrusion or bud starts to form on the parent cell's surface, which will eventually become the new offspring.

DNA Replication: The parent cell duplicates its genetic material and places one copy in the budding bud.

Bud Separation: As the bud grows, it may remain attached to the parent cell for some time. Eventually, the bud becomes large enough to separate from the parent cell entirely.

Independent Growth: The separated bud now becomes an independent bacterial cell, which will continue to grow and undergo further cycles of budding.

Repetition: The process of budding can be repeated in the newly formed cell, allowing for rapid population growth under favorable conditions.

Spore Formation

Bacterial spores are specialized, dormant structures that allow certain bacteria to survive in harsh conditions, such as extreme temperatures, desiccation, and exposure to chemicals or radiation. Spores are highly resistant to environmental stresses, and they can remain viable for extended periods until conditions become favorable for germination.

The process of spore formation in bacteria is called sporulation, and it typically involves several stages:

Initiation: When environmental conditions become unfavorable (e.g., nutrient depletion or other stressors), the bacterial cell initiates the sporulation process.

Replication and Asymmetric Division: The bacterial cell replicates its genetic material, resulting in two DNA copies. The cell then undergoes asymmetric division, creating a larger mother cell and a smaller forespore.

Spore Formation: The mother cell engulfs the forespore, creating a double membrane-bound structure. This eventually becomes the spore, which is now surrounded by a durable spore coat.

Maturation: The spore matures and dehydrates, leading to its increased resistance to harsh conditions.

Release: Once the spore is fully developed, the mother cell lyses, releasing the spore into the environment.

Germination: When conditions become favorable again, the spore germinates, breaking the spore coat, and the viable bacterial cell emerges, resuming its normal vegetative growth and metabolic activities.

Cysts

Cysts are the resting structure under adverse conditions and are created when extra layers are deposited around the mother cell. The mother cell returns to its typical behavior when the circumstances are favorable. The dormant or inactive stage of the mother bacterial cell is called a cyst. When an additional layer is deposited around the mother wall, cysts are created. Poly (3-hydroxybutyrate) (PHB) granules, which are crucial carbon and energy storage molecules in many prokaryotes and enable cell survival in the absence of appropriate carbon sources, are used by bacteria in such unfavorable conditions. . The cell's metabolic process has slowed at this point. When bacteria find a favourable environment, they use a process called excystation to break down the cyst's wall and germinate to generate a new bacterium. Cysts' primary job is to protect the body from harmful environmental changes.

ii. Bacterial recombination

Conjugation

Conjugation is a form of genetic exchange between bacteria that is often referred to as "bacterial sex." However, it's important to clarify that while conjugation involves the transfer of genetic material between bacterial cells, it is still considered a form of asexual reproduction because it does not involve the fusion of gametes or the creation of genetically unique offspring. In bacterial conjugation, the transfer of genes is directional, from a donor to a recipient. The donor "male" has a fertility factor (F+) that is itself heritable. Recipient females do not have the F factor and are F-. Bacteria that have the F factor make the pili needed for conjugation. It was first discovered in *Escherichia coli* by Lederberg and Tatum (1946). They found that two different types of auxotroph (nutritional mutants) grown together on minimal medium produced an occasional prototroph (wild type). Cell contact was required for this change. Anderson (1957) observed conjugation between two such bacteria under electron microscope. Conjugation was later reported in a number of other bacteria.

The process typically involves the following steps:

Donor Cell Preparation: The donor bacterial cell contains a conjugative plasmid (F+), which carries the genes responsible for the conjugation process. These genes encode the necessary structures, such as pilus (a tube-like appendage), that facilitate the transfer of genetic material.

Connection: The donor cell extends a thin, hair-like structure called a pilus towards the recipient cell and establishes physical contact.

Transfer: The plasmid (F+) is then transferred from the donor cell to the recipient cell through the pilus, moving from one cell to the other.

Incorporation: Once inside the recipient cell, the transferred genetic material (plasmid) may be integrated into the recipient's genome or exist as an independent replicating element, depending on the specific characteristics of the plasmid.

Conjugation is an essential mechanism for the horizontal transfer of genes among bacteria. It allows the spread of beneficial traits, such as antibiotic resistance, virulence factors, and metabolic capabilities, between different bacterial species or strains. This pro

Transformation

It is the absorption of DNA segment from the surrounding medium by a living bacterium. The phenomenon was discovered by Griffith in 1928. Its mechanism was worked out by Avery (1944). Transformation is a process by which a bacterium takes up foreign genetic material (usually in the form of DNA) from its surroundings and incorporates it into its own genome. This process is critical for bacterial evolution as it allows the acquisition of new traits that can confer advantages such as antibiotic resistance or the ability to utilize different nutrients.

The process of transformation involves the following steps:

Uptake of DNA: Bacteria can take up external DNA fragments that are released into their environment through the lysis (breaking open) of other bacterial cells or the release of DNA from decaying organic matter. The DNA can be from a closely related species (homologous transformation) or from unrelated species (heterologous transformation).

Integration of DNA: Once taken up, the foreign DNA must be integrated into the recipient bacterium's genome. This process often involves homologous recombination, where segments of the incoming DNA replace the corresponding segments in the recipient's chromosome.

Expression of new genes: After integration, the transformed bacterium may express the new genes present in the acquired DNA. These genes can code for different proteins and enzymes that provide the bacterium with new capabilities or traits.

It's important to note that not all bacteria are naturally competent for transformation (i.e., capable of taking up DNA from their environment). Transformation occurs more readily in certain bacterial species or strains than others. In some cases, researchers can induce competence artificially through laboratory techniques.

Transduction:

It is the transfer of foreign genes by means of viruses. Transduction was first discovered by Zinder and his teacher Lederberg (1952) in *Salmonella typhimurium*. Transduction is one of the processes of horizontal gene transfer in bacteria. It involves the transfer of genetic material from one bacterium to another with the help of bacteriophages, which are viruses that infect and replicate within bacterial cells. The process of transduction can be summarized as follows: These are the two main types of transduction processes observed in bacterial genetics.

Generalized and specialized transduction are two mechanisms by which bacterial DNA can be transferred between bacterial cells. These processes often involve bacteriophages, which are viruses that infect bacteria.

1. Generalized Transduction:

» In generalized transduction, any bacterial DNA can be transferred from one bacterium to another by a bacteriophage during the lytic cycle of viral replication.

» During the lytic cycle, the bacteriophage injects its DNA into the bacterial cell and hijacks the host's cellular machinery to replicate itself.

- Sometimes, mistakes occur during the packaging of newly synthesized phage DNA into the viral capsids. Instead of packaging viral DNA, fragments of bacterial DNA are mistakenly packaged.
- When the phage infects another bacterial cell, it injects this packaged DNA, which can then integrate into the recipient cell's genome via recombination.
- Generalized transduction can transfer any gene from the donor bacterium to the recipient, regardless of its location on the bacterial chromosome or any plasmids it may carry.

2. Specialized Transduction:

- Specialized transduction, in contrast, involves the transfer of specific bacterial genes located near the integration site of a temperate bacteriophage.
- Temperate phages can undergo both the lytic and lysogenic cycles. During the lysogenic cycle, the phage integrates its DNA into the bacterial chromosome.
- Occasionally, when the prophage (integrated phage DNA) is excised from the bacterial chromosome to initiate the lytic cycle, adjacent bacterial DNA may be mistakenly excised along with it.
- This bacterial DNA is then packaged into the phage capsid along with the viral DNA. When this phage infects another bacterium, it injects both viral and bacterial DNA into the recipient cell.
- Specialized transduction is limited to transferring genes that are close to the site of prophage integration.

Both generalized and specialized transduction contribute to horizontal gene transfer among bacteria, allowing for the spread of antibiotic resistance genes, virulence factors, and other beneficial traits.

Infection: A bacteriophage infects a bacterial cell, injecting its genetic material (DNA or RNA) into the host cell.

Replication: Inside the bacterial cell, the phage uses the cellular machinery to produce multiple copies of itself and replicate its own genetic material.

Mistaken Packaging: During the assembly of new phage particles, occasionally, some bacterial DNA may mistakenly get packaged into the viral capsid instead of the phage's own genetic material.

Transfer: The newly formed phage particles, carrying both phage DNA and bacterial DNA, are released when the host bacterial cell lyses (bursts).

Infection of another cell: These phage particles can then infect other bacterial cells. During this process, the bacterial DNA that was mistakenly packaged can be introduced into the new host bacterium.

Incorporation: If the introduced bacterial DNA is compatible with the recipient cell, it can be integrated into the host bacterium's genome through homologous recombination, effectively transferring new genetic traits to the recipient cell.

Transduction allows bacteria to acquire new genetic information, including potentially beneficial traits, from other bacteria within the same environment. It is one of the mechanisms responsible for bacterial diversity and adaptation to changing environmental conditions.

1.3.2. Microbial growth kinetics

The study of bacterial growth kinetics involves examining the rates and patterns of bacterial cell division and population growth under different conditions. Several factors influence bacterial growth, including environmental conditions (e.g., temperature, pH, nutrient availability), genetic characteristics of the bacteria, and interactions with other organisms in their surroundings (Maier and Pepper, 2015). Bacterial growth kinetics can be described using mathematical models that help researchers predict and understand how bacterial populations will change over time. The study of growth kinetics provides valuable insights into bacterial physiology, metabolism, and how these microorganisms respond to changes in their environment.

Lag phase

The lag phase of microorganisms is the initial period of slow or no growth that occurs when they are introduced into a new environment or a fresh growth medium. During this phase, microorganisms are adjusting to the new conditions and preparing for rapid growth in the subsequent exponential phase. The lag phase is an essential part of microbial growth dynamics and has several key characteristics:

Cellular Adaptation: During the lag phase, microorganisms are undergoing various physiological changes to adapt to the new environment. They might be synthesizing specific enzymes, transport proteins, and other molecules needed to utilize the available nutrients effectively.

Repair and Recovery: If the microorganisms experienced stress or damage during transfer to the new environment, they may use the lag phase to repair cellular components and recover from any adverse conditions.

Metabolic Reorganization: Microorganisms often need to adjust their metabolic pathways to make the best use of the nutrients available in the new growth medium.

Length of the Lag Phase: The duration of the lag phase varies depending on the species of microorganism, the complexity of the growth medium, the temperature, pH, and other environmental factors. It can range from a few minutes to several hours or even days.

Growth Heterogeneity: In a population of microorganisms, individual cells may have different lag times due to their varying physiological states or conditions experienced before introduction into the new environment.

The lag phase is followed by the exponential phase (logarithmic phase), during which microorganisms multiply rapidly and their population increases exponentially. The exponential phase is characterized by a constant and balanced growth rate, given that the environmental conditions remain favorable and nutrients are abundant. The length and characteristics of the lag phase can be influenced by various factors, such as the age and condition of the inoculum, the composition of the growth medium, and the presence of growth factors or inhibitors. Researchers and industrial operators often study and manipulate the lag phase to optimize microbial growth and enhance production in biotechnological and industrial processes.

Logarithmic (Exponential) Phase

The logarithmic phase, also known as the exponential phase or log phase, is a critical stage in the growth of microorganisms when they experience rapid and balanced growth. During this phase, the number of microorganisms increases exponentially over time as they divide and multiply at their maximum growth rate under favorable environmental conditions.

Key characteristics of the logarithmic phase of microorganisms include:

Rapid Growth: In the exponential phase, the microorganisms actively divide, and the population size doubles at regular intervals. The growth rate is determined by the generation time, which is the time it takes for the population to double in number.

Balanced Growth: During this phase, the rate of cell division equals the rate of cell death, resulting in a balanced growth rate. The microorganisms are utilizing available nutrients efficiently, and waste products are not yet accumulating at inhibitory levels.

Nutrient Abundance: The exponential phase requires an abundant supply of nutrients, a suitable growth medium, optimal temperature, pH, and other favorable environmental conditions. As long as these factors are maintained, the microorganisms continue to multiply at a consistent rate.

Stability: The logarithmic phase is relatively stable as long as the environmental conditions remain constant. The growth curve shows a straight line on a semilogarithmic graph, reflecting the exponential increase in the number of microorganisms.

Population Doubling: Since the growth is exponential, the number of microorganisms in the population doubles at regular intervals. For example, if the generation time is 20 minutes, the population will double every 20 minutes.

High Metabolic Activity: During the exponential phase, the microorganisms have high metabolic activity, and biosynthetic processes are operating at their peak to support cell division and biomass production.

The exponential phase is a crucial period in various applications, including biotechnology, fermentation, and industrial microbiology. It is desirable in these fields to maintain microorganisms in the logarithmic phase to achieve high biomass or product yields efficiently. However, the exponential phase cannot be sustained indefinitely. Eventually, as the microorganisms consume nutrients and produce waste products, the growth rate will slow down, leading to the stationary phase, where the rate of cell division equals the rate of cell death. Understanding the exponential phase's duration and characteristics is essential for optimizing microbial growth and controlling microbial populations in various applications.

Stationary Phase

The stationary phase of microorganisms is the third phase in the typical growth curve of a microbial population. During this phase, the growth rate of the microorganisms slows down, and the number of viable cells remains relatively constant over time. The stationary phase occurs after the exponential phase and is followed by the death phase.

Key characteristics of the stationary phase of microorganisms include:

Equilibrium: In the stationary phase, the growth rate of the microorganisms becomes balanced, and the number of cells dividing is equal to the number of cells dying. As a result, the total population size remains relatively constant or fluctuates only slightly.

Nutrient Depletion: The stationary phase is reached when the growth medium's nutrients become depleted or limited. As the microorganisms consume available nutrients, the environment becomes less favorable for exponential growth.

Accumulation of Waste Products: During the stationary phase, waste products from the metabolism of the microorganisms can accumulate to inhibitory levels. These waste products can include organic acids, alcohols, and other metabolic byproducts.

Reduced Metabolic Activity: With limited nutrients and increased waste product accumulation, the metabolic activity of the microorganisms decreases compared to the exponential phase. Cell division slows down, and some cells may enter a dormant or non-replicative state.

Sporulation or Differentiation: In some microorganisms, the stationary phase can trigger sporulation or differentiation into specialized forms that are more resistant to adverse conditions. This is a survival strategy to cope with nutrient limitation and other stresses.

Longer Duration: The stationary phase can last for an extended period, depending on the availability of nutrients and the accumulation of waste products. The length of the stationary phase varies depending on the species of microorganism and the specific growth conditions.

The stationary phase is a significant consideration in various fields, including biotechnology, microbiology, and food preservation. In biotechnological processes, the stationary phase can be a challenge as it can lead to reduced productivity and product yields. Strategies to extend the exponential phase and delay entry into the stationary phase are often employed to maximize product output. In microbial fermentation processes, reaching the stationary phase is usually desired, as it indicates that the desired product has been synthesized and accumulated. However, in cases where continuous growth is needed, the stationary phase can be problematic.

Overall, the stationary phase represents a period of stability in the microbial population, where the growth rate levels off due to nutrient depletion and waste product accumulation. Understanding the characteristics of the stationary phase is essential for optimizing microbial growth and maintaining microbial populations in various applications.

Death phase

The death phase of microorganisms is the final stage in the typical growth curve of a microbial population. During this phase, the number of viable cells in the population decreases over time as the microorganisms dies off. The death phase follows the stationary phase and occurs when the environmental conditions become increasingly unfavorable for the survival and growth of the microorganisms.

Key characteristics of the death phase of microorganisms include:

Decreasing Viability: In the death phase, the number of viable cells decreases steadily. Cell death can occur due to various factors, including nutrient exhaustion, the accumulation of toxic waste products, depletion of essential cofactors, and adverse environmental conditions.

Nutrient Depletion: The main cause of the death phase is the depletion of nutrients in the growth medium. As the microorganisms consume all available nutrients during the exponential and stationary phases, they are no longer able to sustain their metabolic processes and cell division.

Waste Product Toxicity: Accumulation of toxic waste products, such as organic acids and metabolic byproducts, during the stationary phase can become inhibitory to the microorganisms, leading to cell damage and death.

Declining Metabolic Activity: With limited nutrients and the presence of toxic waste products, the metabolic activity of the microorganism's decreases significantly compared to the exponential and stationary phases.

Longer Duration: The death phase can last for an extended period, depending on the initial population size, the specific microorganism, and the environmental conditions. Some microorganisms may enter a viable but non-culturable state during this phase, making it difficult to detect their presence by conventional culturing methods.

Survival Strategies: Some microorganisms may employ survival strategies during the death phase, such as the formation of endospores (sporulation) or protective structures, to endure adverse conditions and potentially revive when conditions become more favorable.

The death phase is a natural consequence of the limited resources available in the growth environment. In natural environments and during batch fermentation processes, the death phase occurs as the microorganisms exhaust the available nutrients and encounter adverse conditions. In continuous culture systems or bioreactors, the death phase can be minimized or avoided by providing a continuous supply of fresh nutrients, thus maintaining the microorganisms in the exponential or stationary phase.

It is essential to consider these growth kinetics when working with microorganisms in various applications. For example, in biotechnology and industrial processes, maintaining microorganisms in the exponential phase is often desired to achieve high biomass or product yields. On the other hand, in food safety and preservation, understanding the death phase is critical to prevent the growth of spoilage or pathogenic microorganisms.

Microbial growth kinetics can be studied using various techniques, including measuring the increase in cell numbers over time using cell counting methods, monitoring changes in biomass, or analyzing changes in metabolic products during different growth phases. Understanding the growth behavior of microorganisms is fundamental to control their proliferation in diverse applications and contexts.

Chapter - 3

In-situ Bioremediation Process

3.1. Introduction

Bioremediation is basically a result of metabolic activity of living microorganism of soil and aquatic ecosystem. Due to the metabolic activity the organism utilises the different organic or inorganic pollutants from their habitat/media as nutrient consequently they reduces the pollution load from the specific site. Therefore, bioremediation relies on the metabolic capabilities of specific microorganisms, such as bacteria, fungi, protozoans or actinomycetes, algae etc. to degrade or transform hazardous substances into less harmful forms. These microorganisms can break down a wide range of contaminants, including petroleum hydrocarbons, solvents, pesticides, and heavy metals for their accumulation as their nutrient by releasing the extracellular enzymes.

In-situ bioremediation is a remediation technique used to clean up environmental contamination by harnessing the natural abilities of microorganisms. The term "in-situ" means that the process takes place at the site of contamination itself, without the need for excavation or transport of contaminated materials. Instead, the contaminants are treated directly in their original location. The process of in-situ bioremediation involves creating conditions that promote the growth and activity of the appropriate microorganisms.

3.2. Types of In-situ Bioremediation

The in-situ bioremediation technique may be dived broadly into two categories *natural attenuation* and *enhanced*. The enhanced in-situ bioremediation technique can be further sub-categorized depending on their process (Sharma, 2019). This can be elaborated as below:

DOI: 10.1201/9781003787549-3

One of the advantages of in-situ bioremediation is its ability to treat contaminants on-site, reducing the need for costly and disruptive excavation and transportation of contaminated materials. It can be applied to a variety of contaminated environments, including soil, groundwater, and surface water. Additionally, in-situ bioremediation is often considered an environment friendly approach since it utilizes natural processes and can be less energy-intensive compared to other remediation methods.

However, in-situ bioremediation does have its limitations. The success of the process depends on various factors, including the type and concentration of contaminants, site conditions, and the presence of suitable microbial populations. Some contaminants may be more resistant to biodegradation or may require longer treatment periods. It is also important to consider the potential for the release of by-products during the bioremediation process and their potential impact on the environment.

3.2.1. Bioventing

Bioventing is an in-situ remediation technology that uses indigenous microorganisms to biodegrade organic constituents adsorbed to soils in the unsaturated zone. Soils in the capillary fringe and the saturated zone are not affected. In bioventing, the activity of the indigenous bacteria is enhanced by inducing air (or oxygen) flow into the unsaturated zone (using extraction or injection wells) and, if necessary, by adding nutrients.

When extraction wells are used for bioventing, the process is similar to soil vapor extraction (SVE). However, while SVE removes constituents primarily through volatilization, bioventing systems promote biodegradation of constituents and minimize volatilization (generally by using lower air flow rates than for SVE). In practice, some degree of volatilization and biodegradation occurs when either SVE or bioventing is used.

This technique involves controlled stimulation of airflow by delivering oxygen to unsaturated (vadose) zone in order to increase bioremediation, by increasing activities of indigenous microbes. In bioventing, amendments are made by adding nutrients and moisture to enhance bioremediation with the ultimate goal being to achieve microbial transformation of pollutants to a harmless state.

Soil normally contains large numbers of diverse microorganisms including bacteria, algae, fungi, protozoa, and actinomycetes. In well-aerated soils, which are most appropriate for bioventing, these organisms are generally aerobic. Of these organisms, the bacteria are the most numerous and biochemically active group, particularly at low oxygen levels. Bacteria require a carbon source for cell growth and an energy source to sustain metabolic functions required for growth. Nutrients, including nitrogen and phosphorus, are also required for cell growth. Hydrocarbon-degrading aerobic bacteria use oxygen to metabolize organic material to yield carbon dioxide and water, a process commonly referred to as aerobic respiration. To degrade large amounts of petroleum a hydrocarbon, a substantial bacterial population is required which, in turn, requires oxygen for both the metabolic process and the growth of the bacterial mass itself. Approximately 3 to 3.5 pounds of oxygen are needed to degrade one pound of petroleum product.

Unlike bioventing that relies on enhancing microbial degradation process at the vadose zone by moderate air injection, soil vapour extraction (SVE) maximizes volatile organic compound volatilization via vapour extraction.

The most important factors that control the effectiveness of bioventing are (Smith et al., 2023):

- The permeability of the petroleum-contaminated soils. This will determine the rate at which oxygen can be supplied to the hydrocarbon-degrading microorganisms found in the subsurface.
- The biodegradability of the petroleum constituents. This will determine both the rate at which and the degree to which the constituents will be metabolized by microorganisms.
- During on-site field testing, getting equivalent findings to those obtained in laboratory studies is not always possible due to various environmental conditions and differences in the unsaturated zone into which air is pumped; as a result, treatment time with bioventing may be extended.
- Apparently, high airflow rates cause the transfer of volatile organic compounds to the soil vapour phase, necessitating off-gas treatment of the resultant gases before release into the atmosphere.
- This particular challenge can be resolved by combining bioventing and biotrickling filter techniques to reduce both contaminant and outlet gas emission levels; thus reducing the extended treatment time associated with bioventing alone.

Application of Bioventing

- Bioventing is used to reduce a wide range of petrochemical compounds, including gasoline, fuel oil and bitumen.
- The constituents of these compounds are measured together as total petroleum hydrocarbons (TPH).
- Site Assessment

3.2.2. Biosparging

This approach, like bioventing, involves injecting air into the soil subsurface to encourage microbial activity and enhance pollution removal from polluted locations. However, unlike bioventing, air is introduced into the saturated zone, which can allow volatile organic compounds to rise to the unsaturated zone and enhance biodegradation (Bell et al., 2022). The efficiency of biosparging is determined by two key factors: soil permeability, which influences pollutant bioavailability to microorganisms, and pollutant biodegradability.Biosparging is frequently used to treat aquifers contaminated by petroleum compounds, particularly diesel and kerosene.

Application of Biosparging technology

- Biosparging has been widely applied to the treatment of aquifers polluted by oil derivatives, mainly kerosene and diesel, which have good biodegradation of the BTEX (benzene, toluene, ethylbenzene and xylene) group and naphtalenes.
- The effectiveness of biosparging depends on soil permeability that determines the availability of pollutants to microorganisms as well as the pollutant biodegradability
- This technique is suitable for onsite (in situ) remediation of both soil and groundwater and is the best option for mineral oil removal there is. Even large amounts of mineral oil can be removed with this method (up to ten thousands of mg kg-1 TDS).

3.2.3. Bioslurping

Bioslurping is combination of two remedial technologies i.e. bioventing and vacuum-enhanced free-product recovery. Bioventing encourages the aerobic bioremediation of hydrocarbonimpacted soils (Sravya and Sangeetha, 2022). While, vacuum-enhanced free-product recovery removes light non-aqueous phase liquids (LNAPLs) from the capillary fringe and the water table below ground.

Bioslurping combines aspects of bioventing and free product recovery to address two distinct contamination media by recovering free product while also bioremediating the soil's vadose zone. Bioslurping can increase free-product recovery without draining large amounts of groundwater. Vacuum-enhanced pumping lifts the LNAPL off the water table and removes it from the capillary fringe, reducing water table changes and helping to prevent a smear zone. The goal of bioventing is to reduce groundwater and soil gas emissions into the environment. When free-product removal is complete, bioslurping is converted into a standard bioventing system to complete the remediation process. Depending on site conditions, operating and maintenance times can range from a few months to years.

Bioslurping has been used to treat soils contaminated with petroleum hydrocarbons. It is a cost-effective method that removes LNAPLs while also remediating the soil in the vadose zone. Bioslurping can also be used in areas with a deep subsurface water table (>30 feet).

Limitations:

» Bioslurping is less efficient in low-permeability soils.

» Low soil moisture content can impede biodegradation and reduce the effectiveness of the bioventing process, which causes soils to dry out.

» Aerobic biodegradation of chlorinated chemicals may not be effective unless a co-metabolite is present.

» Low temperatures limit cleanup.

» The bioslurper system's off-gas requires treatment before it may be discharged. However, due to lowering fuel prices, off-gas treatment can only be carried out quickly after the system is started.

» At certain sites, bioslurper systems can remove large volumes of water that may require treating before discharge, depending on the concentration of contaminants present in the process water.

» Due to the fact that fuel, water and air are removed from the subsurface in a single stream, mixing of the phases almost always occurs.

» Such mixtures require special oil/water separators or treatment before the treated water can be discharged.

3.2.4. Biostimulation

Biostimulation is a process of in-situ bioremediation technology where the bioremediation activity is initiated and enhanced by addition of nutrient to provide the adequate condition to indigenous microbes for the degradation of any pollutants particular polluted sites. Biostimulation involves the modification of the environment to stimulate existing bacteria communities capable of bioremediation. This can be done by addition of various forms of rate limiting nutrients and electron acceptors, such as phosphorus, nitrogen, oxygen, or carbon (e.g. in the form of molasses). Hence, biostimulation may be defined as a process of natural remediation that can be improved pollutant degradation by optimizing the condition such as aeration, addition of nutrients, pH and temperature control. It has been reported by various researcher that biostimulation can be considered as an appropriate remediation technique for petroleum pollutants removal in soil and requires the evaluation of both the intrinsic degradation capacities of the autochthonous microflora and the environmental parameters involved in the kinetics of the in situ process. Thus, biostimulation is a remediation technique that is highly efficient, cost effective and eco-friendly in nature. Biostimulation refers to the addition of rate limiting nutrients like phosphorus, nitrogen, oxygen, electron donors to severely polluted sites to stimulate the existing bacteria to degrade the hazardous and toxic contaminants.

The key benefit of biostimulation is that bioremediation will be carried out by already extant native microorganisms that are well suited to the subsurface environment and are evenly distributed across the subsurface. The primary challenge is that the delivery of additives in a manner that allows the additives to be readily available to subsurface microorganisms is based on the local geology of the subsurface, i.e. the characteristics of soil properties. Tight, impermeable subsurface lithology (tight clays or other fine-grained material) make it difficult to spread additives throughout the affected area. Fractures in the subsurface create preferential pathways for additives to follow, hindering a uniform distribution of additives. Addition of nutrients might also promote the growth of heterotrophic microorganisms which are not innate degraders of Total Petroleum Hydrocarbon thereby creating a competition between the resident micro flora.

The functional property of any nutrient or chemical compound in soil (sediments) or groundwater will be fundamentally different with the microbial activities due to their physical properties i.e. the nature of the soil will be different than water. Even the smallest bacterium has different adsorption properties from chemicals. Clayey soils, for example, have extremely low porosity and may prevent bacteria from penetrating. These clays may also bind the microbes that are added, for example, via cationic bridges involving divalent metals and the net negative charge on the bacteria's and clay's surfaces. In some soils, injected inorganic nutrients can precipitate metals, expand clays, and modify redox potentials and conductivity, all of which have a significant impact on groundwater flow and environmental biogeochemistry. Indeed, bacterial plugging of subsurface formations has been employed successfully to increase oil recovery in oil reservoirs.

Thus, biostimulation is dependent on indigenous organisms and thus requires their presence, as well as the ability to alter the environment in such a way that the desired bioremediation effect is achieved. Because of the presence of a potential bacterial colony at the location and the necessary food requirements for their healthy growth. In most terrestrial subsurface settings, indigenous organisms have been exposed to the contamination for long periods of time and have adapted or even evolved spontaneously. Many pollutants, particularly organic chemicals, are naturally occurring or have natural analogs in the environment. A terrestrial subsurface ecosystem with a lack of organisms capable of degrading or transforming any pollutant is uncommon. Indeed, even pristine environments have bacteria with an increasing number of plasmids with sediment depth in response to increasing recalcitrance of the organics present.

3.2.5. Bioaugmentation

Bioaugmentation is a type of in-situ bioremediation technique for improvement of pollutant degradation by addition of potential microbial strain on the pre-existing microbial community of any polluted site (Mutter, 2023). Bioaugmentation consists of the addition (augmentation) of specialized microbial cultures, which are typically grown separately under well-defined conditions, to perform a specific remediation task in a given environment.

The utilization of bioaugmentation provides advancement in the fields of microbial ecology and biology, immobilization, and bioreactor design. Bioaugmentation is commonly used in municipal wastewater treatment to restart activated sludge bioreactors. Most cultures available contain microbial cultures, already containing all necessary microorganisms (*B. licheniformis*, *B. thuringiensis*, *P. polymyxa*, *B. stearothermophilus*, *Penicillium* sp., *Aspergillus* sp., *Flavobacterium*, *Arthrobacte*, *Pseudomonas*, *Streptomyces*, *Saccharomyces*, etc.).

Two different bioaugmentation approaches have been developed. One approach involves injecting microorganisms with the required catabolic potential to supplement or replace the local microbe population. In this instance, the chosen bacteria or consortia may survive and outcompete native microorganisms, and they occupy a specific metabolic niche in the contaminated environment. The second bioaugmentation method involves adding a large number of cells that serve as biocatalysts for a short period of time before becoming inactive or dying. In this instance, the injected microorganisms are unable to establish due to the intrinsic abiotic and biological stress of the new environment. These include temperature swings or extremes, pH, water activity, nutrient deficiency, harmful pollutant concentrations, and competition with indigenous microorganisms. In such instances, repeated biomass re-injections are required over time since the infected cells are unable to thrive in situ. Biostimulation (the addition of nutrients and other stimulatory substrates and/or electron acceptors as appropriate) is commonly used in conjunction with bioaugmentation to improve the survival of the added cells and/or optimize their metabolic capabilities, resulting in robust long-term biodegradation efficiency.

Bioaugmentation in wastewater activated sludge systems is relatively simple because the added microorganisms can be easily mixed in the reactor and reaction conditions can be adjusted to improve their survival and performance (for example, improved flocculation and settling of biomass, faster nitrification, and enhanced degradation of recalcitrant compounds). Bioaugmentation of aquifers, on the other hand, is more difficult and must account for challenges such as the added strains' survival, distribution throughout the contaminated zone (which is frequently influenced by biomass growth near the injection wells), and low concentrations of nutrients and target contaminants that serve as substrates for the added microorganisms. Even so, the benefits of bioaugmentation have been demonstrated in field trials for a wide variety of recalcitrant contaminants (> Table 1) and thus has become of increasing commercial interest over the past decade. For example, the Green Pages (www.eco-web.com) lists 632 companies that match the keyword "bio-augmentation." Nonetheless, the success of bioaugmentation cannot always be demonstrated due to the difficulty to assess whether the inoculated cells are responsible for the increased removal and degradation of the target contaminant as compared to non-bioaugmented controls. Bioaugmentation produces shorter acclimation times and faster degradation, frequently with less undesirable byproducts, while a capable indigenous consortium may eventually form at some contaminated areas (leading to the eventual breakdown of the targeted pollutants). Therefore, depending on native microbial activity, which could have a comparatively long lag phase, might not be suitable if a quick response is required.

Several factors can influence the success and effectiveness of bioaugmentation (Ma et al., 2022). Here are some key factors:

1. Microbial Strain Selection: The selection of appropriate microbial strains is crucial for bioaugmentation. The selected strains should be capable of performing the desired function, such as degrading specific pollutants or enhancing nutrient cycling. Factors to consider include the strain's metabolic capabilities, environmental compatibility, growth rate, and resistance to adverse conditions.
2. Environmental Conditions: The environmental conditions at the site of bioaugmentation play a significant role in its success. Factors such as temperature, pH, moisture, oxygen availability, and nutrient availability can affect the growth and activity of the introduced microorganisms. It is important to assess and optimize these conditions to ensure the survival and effectiveness of the augmented microbial population.
3. Nutrient Availability: Microorganisms require essential nutrients for growth and metabolism. Adequate availability of nutrients, such as carbon, nitrogen, phosphorus, and trace elements, is important for the successful establishment and activity of the augmented microbial population. If the nutrient levels are insufficient, the growth and function of the introduced microorganisms may be limited.
4. Compatibility with Indigenous Microorganisms: Indigenous microorganisms already present in the environment can interact with the introduced microbial strains. Compatibility between the augmented and indigenous microbial populations is important to avoid competition or negative interactions. Assessing the composition and activity of the native microbial community and selecting compatible strains can help enhance the success of bioaugmentation.
5. Bioavailability of Pollutants: The bioavailability of pollutants, such as organic compounds or heavy metals, can influence the efficiency of bioaugmentation. Some pollutants may be tightly bound to soil particles or exist in complex chemical forms that are less accessible to microorganisms. Enhancing the bioavailability of pollutants through techniques like biostimulation (adding nutrients) or bioremediation strategies (using specific strains) can improve the effectiveness of bioaugmentation.
6. Application Strategy and Monitoring: The method and timing of introducing the microbial strains are important considerations. Factors such as the concentration and frequency of application, inoculation methods, and monitoring protocols can affect the success of bioaugmentation. Regular monitoring of microbial activity, pollutant levels, and environmental parameters can provide feedback

and help optimize the bioaugmentation process.

7. Site-specific Factors: Each bioaugmentation project is unique, and site-specific factors can influence the outcome. These factors may include the presence of inhibitors or toxic substances, the physical characteristics of the site (e.g., soil texture, hydrology), the presence of vegetation or other biota, and regulatory considerations. A comprehensive understanding of the site-specific factors is essential for successful bioaugmentation.

Overall, careful consideration of these factors and proper planning and implementation of bioaugmentation strategies can maximize the chances of success and achieve the desired outcomes in various environmental applications.

Application of Bioaugmentation

» Bioaugmentation is commonly used in municipal waste water treatment to restart activated sludge bioreactors.

» Most cultures available contain microbial cultures, already containing all necessary microorganisms *B. licheniformis*, *B. thuringiensis*, *stearothermophilus*, *Penicillium sp.*, *Aspergillus sp.*, *Flavobacterium*, *Athrobacter*, *Pseudomonas*, *Streptomyces*, *Saccharomyces*, etc.)

» Activated sludge systems are generally based on microorganisms like bacteria, protozoa, nematodes, rotifers, and fungi, which are capable of degrading biodegradable organic matter.

» There are many positive outcomes from the use of bioaugmentation, such as the improvement in efficiency and speed of the process of breaking down substances and the reduction of toxic particles in an area.

Chapter - 4

Ex-situ Bioremediation Process

4.1. Introduction

If we have any polluted site and we remove the pollutants from the original environment and let bioremediation happen off-site, we call that ex-situ bioremediation. When the polluted site material is removed from the environment, it can be put into bioreactors, large vessels where the contaminated material can be monitored and conditions for bioremediation can be controlled. Biological organisms typically require optimum conditions where they can perform best. In bioreactors we can control the mixing rate, temperature, pH, and nutrient levels as per requirement of microorganism to breakdown the targeted pollutants. Landfarming involves spreading contaminated soil into a lined bed (to prevent leaching) and periodically applying nutrients and mixing the soil to boost biological activity. Biopiling places the contaminated soil into piles that are well aerated and nutrients are added to speed up bioremediation. In all cases, the contaminant levels are monitored to verify that bioremediation is taking place and steps are taken to ensure that contaminated material stays out of contact with the environment.

The formation of various organo-metallic complex or other complex of organic and inorganic compounds is assumed to be formed either by an attachment of the compounds to reactive sites on the surface of the organic colloids or by incorporation of the compounds into structure of humic or fulvic acids that are being produced through microbial action.

There are several organic and inorganic compounds which are discharge from either industrial activities or different natural process and change the original composition of soil, water or air. Such undesirable or unwanted substances are known as either contaminants or pollutants. As we know that soil is very unique part of the

DOI: 10.1201/9781003787549-4

terrestrial ecosystem, its role in the environment as reservoir for the growth of plant and microbial population for the degradation and recycling of various dead biomasses, which become very important component of the environment for maintaining the life on earth. If we see the soil composition it has very heterogeneous constituent and often comprises mineral and organic solids, aqueous and gaseous component. The mineral component of soil often consist of weathering product of rock parent material and secondary mineral such as clay mineral oxides of iron (Fe), aluminum (Al), manganese (Mn) and sometime carbonate such as calcium carbonate ($CaCO_3$). Living organism (mesofauna and microfauna), dead plants and colloidal humus originated by the action of microorganism on the plant litter make up the organic matter component of the soil. These soil components are usually clustered together in the form of aggregates, thus creating a system of interconnected pores of various sizes filled with either water or air in form of soil capillaries or capillaries fringe in unsaturated zone of sub-surface soil. The resulting soil-air-water environment becomes very complex; this is because different soil complex gives rise to different reaction at the subsurface of the soil when anthropogenic chemicals are introduced. In many cases contaminants become bonded to the soil specially to clay or organic materials which become unavailable to the indigenous microbial population for their degradation. This become as a risk. In other situation the added compounds are anthropogenic chemicals may remain unbounded in the soil acidity or redox potential and create adverse effect immediately to existing to flora or fauna. In addition many xenobiotics including the pesticide have been found in soil as a bound residue of soil particles. Most of the complexes formed with two organic or humus fraction of soil, generally they involve in the formation of covalent bond between the compound and humic substances.

The goal of a remediation effort is to limit the extent of contamination at a hazardous waste site, to prevent further deterioration of the environment and to prevent exposure by humans and other life forms to hazardous chemicals. The remedies at a given site vary depending on the properties of the chemicals found at the site, the types of soils, and the depth of contamination, and natural processes that may occur at the site. The criteria for selecting remediation technologies or treatments are:

- Short-term and long-term effectiveness at meeting the remediation goals
- Reduction in the volume of contaminants most effectively
- Reduction in the toxicity of contaminants
- Cost effectiveness.

Other factors affect the choice of remedies-for example, land disposal of hazardous materials is restricted under present environmental regulations. Because a variety of remedies can often control contamination at hazardous waste sites, selecting the appropriate technology or technologies can be a challenge. The following remedy types are arranged from least disruptive, intensive, and costly to most disruptive, intensive, and costly, i.e. (i) Take no action: No remediation action is necessary if the environmental assessment shows that neither the environment nor people are in danger. (ii) Institutional Controls (IC): To prevent access to hazardous regions, institutional controls are employed at polluted locations. Fencing off a contaminated area or forbidding the drilling of wells close to toxic aquifers are two instances of institutional restrictions. (iii) Monitored Natural Attenuation: Sometimes contaminants are cleaned up by nature. Examples include natural bioremediation, photolysis (the breakdown of pollutants into harmless elements by sunshine), and chemical reactions (such hydrolysis). Additionally, pollutants have the ability to "stick" (or sorb) to soil or other solid particles, which restricts their mobility within the soil. (iv) Containment of toxins: By restricting the ways that people, animals, and the environment can come into contact with the toxins, the danger of environmental contamination can be decreased. Techniques include stabilizing or solidifying contaminants in place, building physical barriers, and covering soils with clean material. Contaminated groundwater is more challenging to contain and may require complex, very costly pumping and treatment systems. (v) Destruction of contaminants: If the byproducts are not harmful, destroying the contaminants may eliminate the risk they cause. Treatments can be in situ (in place) or ex situ (after excavation of the soil or pumping of the ground water from the aquifer). Technologies include phytoremediation, bioremediation, Fenton's Reagent and ozonization. (vi) *Removal*: In some cases, the best option may be to physically remove the contaminated soil and move it to a permitted Treatment, Storage and Disposal Facility (TSDF). Contaminants in groundwater can be removed using pump-and-treat technologies involving methods such as activated carbon or ion exchange

But when the contaminants are very toxic and complex, they can leach down to contaminate groundwater or they can create offensive odors around the contaminated area. In such conditions the contaminants have to be removed immediately for the adequate solution. Then either contaminated area may be excavated and transported to treat some other place of the pollutants has to be removed and dumped to nay identified safe places by taking appropriate measures. Then this will be known as ex-situ bioremediation technology.

4.2. Types of ex-situ Bioremediation

The ex-situ bioremediation may be of following types:

a. Land farming
b. Biopiling
c. Bioreactor treatment
d. Composting
e. Constructed wetland treatment process.

4.2.1. Land farming: Landfarming, also known as land treatment, this is simplest bioremediation technique owing to its low cost and less equipment requirement for operation. This technique involves the excavation of the contaminated soil and spreading it on a thin surface. In land farming, one thing is common, polluted soils are usually excavated and/or tilled, but the site of treatment apparently determines the type of bioremediation (Maila and Cloete, 2004). When excavated polluted soil is treated on-site, it can be regarded as in situ; otherwise, it is ex situ as it has more in common with other ex situ bioremediation techniques. Biodegradation of pollutants is stimulated aerobically by tilling or plowing the soil. Nutrients (N, P and K) and minerals are also added to promote the growth of the indigenous species. Thus, landfarming is an ex-situ waste treatment process that is performed in the upper soil zone or in biotreatment cells. Contaminated soils, sediments, or sludges are transported to the landfarming site, incorporated into the soil surface and periodically turned over (tilled) to aerate the mixture. Landfarming commonly uses a clay or composite liner to intercept leaching contaminants and prevent groundwater pollution, however, a liner is not a universal requirement. This technique has been used for years in the management and disposal of drill cuttings, oily sludge and other petroleum refinery wastes. The equipment employed in land farming is typical of that used in agricultural operations. These land farming activities cultivate and enhance microbial degradation of hazardous compounds. As a rule of thumb, the higher the molecular weight (i.e., the more rings within a polycyclic aromatic hydrocarbon), the slower the degradation rate. Also, the more chlorinated or nitrated the compound, the more difficult it is to degrade.

According to The Unites States Environmental Protection Agency (EPA) report on underground storage tanks has to be prepared to assess the degradability and safe disposal without causing adverse effect on groundwater. Hence, before the remediation takes place the site must be prepared by clearing and grading the soil, by installing leachate collection and treatment systems, and also building vapor treatment facilities.

It has been reported that when a pollutant lies <1 m below ground surface, bioremediation might proceed without excavation, while pollutant lying >1.7 m needs to be transported to ground surface for bioremediation to be effectively enhanced. Land farming is usually used for remediation of hydrocarbon-polluted sites including polyaromatic hydrocarbons ; as a result, biodegradation and volatilization (weathering) are the two remediation mechanisms involved in pollutant removal. Land farming system complies with government regulations, and can be used in any climate and location. The construction of a suitable land farming design with an impermeable liner minimizes leaching of pollutant into neighbouring areas during bioremediation operation. Over all, land farming bioremediation technique is very simple to design and implement, requires low capital input and can be used to treat large volume of polluted soil with minimal environmental impact and energy requirement. Although the simplest bioremediation technique, land farming like other ex situ bioremediation techniques has some limitations, which include: large operating space, reduction in microbial activities due to unfavourable environmental conditions, additional cost due to excavation, and reduced efficacy in inorganic pollutant removal (Nasreen and Kalsoom, 2018). Moreover, it is not suitable for treating soil polluted with toxic volatiles due to its design and mechanism of pollutant removal (volatilization), especially in hot (tropical) climate regions. These limitations and several others make land farming based bioremediation time consuming and less efficient compared to other ex situ bioremediation techniques.

Application of Land farming:

» *Oil Spill Cleanup*: Land farming has been successfully used in the cleanup of oil spills on land. The contaminated soil is excavated and spread out in a treatment area. Organic amendments, such as compost or manure, are then applied to the soil surface. Microorganisms present in the soil break down the hydrocarbons in the presence of oxygen, moisture, and nutrients provided by the amendments. Regular monitoring and turning of the soil help to optimize microbial activity and enhance the degradation process.

» *Refinery and Industrial Sites*: Land farming is commonly employed at refinery and industrial sites where soil contamination occurs due to accidental spills, leaks, or improper disposal practices. Excavated contaminated soil is treated in land farming units, where it is mixed with organic amendments. The process helps to break down the hydrocarbons over time, reducing their concentration in the soil.

» *Agricultural and Livestock Operations*: Land farming can also be utilized in agricultural and livestock operations to manage and treat organic waste materials generated on-site. By applying these waste materials onto designated areas of land, the organic matter is broken down by microorganisms, improving soil fertility and reducing potential environmental impacts.

» *Bioremediation of Pesticides*: Land farming has shown promise in the bioremediation of soils contaminated with pesticides. Certain microorganisms have the ability to degrade various pesticide compounds, and land farming provides a suitable environment for their growth and activity. Careful selection of organic amendments and monitoring of soil conditions are necessary for effective pesticide degradation.

» *Landfill Remediation*: Land farming can be used as part of landfill remediation efforts. Excavated soil from contaminated areas within the landfill can be treated using land farming techniques to facilitate the degradation of organic contaminants. This approach helps to reduce the overall environmental impact and facilitates the eventual reuse of the treated soil.

Advantage of Land farming:

» *Cost-effectiveness:* Landfarming is often considered a cost-effective remediation option, especially for large-scale contaminated sites. It typically requires lower capital and operational costs compared to other remediation technologies. The simplicity of the technique and the use of natural processes make it a relatively affordable option for treating petroleum-contaminated soil.

» *Minimal Site Disruption*: Landfarming can be conducted on-site, minimizing the need for excavation and off-site transportation of contaminated soil. This reduces disruption to the surrounding area and avoids the potential risks associated with soil transportation, such as accidents or spills. It is particularly advantageous for sites with limited access or sensitive ecosystems where excavation and transport may not be feasible or desirable.

» *Environmental Compatibility*: Landfarming is an environmentally friendly remediation method as it relies on natural processes to degrade contaminants. It utilizes indigenous soil microorganisms that break down petroleum hydrocarbons through biodegradation. This approach avoids the use of chemicals or invasive techniques, minimizing the generation of waste or by-products and reducing the potential for secondary environmental impacts.

» *Soil Restoration:* In addition to contaminant removal, landfarming can improve soil quality and restore fertility. Through the addition of amendments such as compost, manure, or other organic materials, landfarming enhances soil organic matter content, nutrient availability, and overall soil health. This can promote the growth of beneficial microorganisms, increase soil moisture retention, and enhance the long-term productivity of the treated land.

- *Simplicity and Scalability*: Landfarming is a relatively simple and scalable technique that can be easily implemented and managed. The process involves spreading contaminated soil in a controlled manner, incorporating appropriate amendments, and regularly turning the soil to promote aeration and microbial activity. The simplicity of the method allows for flexibility in the size and configuration of treatment areas, making it suitable for small or large-scale projects.

- *Long-term Monitoring and Management*: Landfarming provides the opportunity for long-term monitoring and management of the remediation process. Regular monitoring of soil parameters, such as contaminant concentrations, moisture content, and temperature, allows for adjustment of management practices to optimize degradation rates. It also enables early identification of any potential issues or challenges that may arise during the treatment process.

- *Public Acceptance*: Landfarming is a well-established and accepted remediation technique for petroleum-contaminated soil. It has a proven track record of success in numerous remediation projects worldwide, which contributes to public confidence in the method. The familiarity and acceptance of landfarming can simplify the permitting and regulatory processes associated with remediation projects.

Limitations of Land farming:

While land farming can be an effective method for remediating soil contamination, it also has certain limitations. Some of the limitations of land farming include:

- *Contaminant Limitations:* Land farming is most effective for the treatment of soil contaminated with petroleum hydrocarbons. It may not be suitable for the remediation of soils contaminated with heavy metals, radioactive materials, or certain persistent organic pollutants. Different contaminants require specific remediation approaches, and land farming may not be the optimal solution for all types of soil contamination.

- *Soil Compatibility:* Land farming is most successful when applied to soils with specific characteristics. Soils that are too sandy or have a low organic matter content may not provide the ideal conditions for microbial activity and biodegradation. Similarly, soils with high clay content may lead to poor drainage and reduced oxygen availability, hindering the degradation process.

- *Time Requirements*: Land farming is a time-consuming process that can take months or even years to achieve the desired level of remediation. The length of time required depends on various factors, including the initial contaminant concentration, soil properties, climate conditions, and the effectiveness of management practices. In situations where a rapid cleanup is necessary, alternative remediation methods may be more suitable.

- *Space Requirements*: Land farming typically requires a significant amount of land for the treatment and storage of contaminated soil. Adequate space is needed to spread out the soil, apply amendments, and allow for regular turning and monitoring. This can be a limitation in areas where land availability is limited or expensive.

- *Regulatory Compliance:* Land farming is subject to regulatory requirements and permits due to its potential environmental impacts. Compliance with regulations and obtaining necessary permits can be a complex and time-consuming process. Failure to meet regulatory standards can lead to delays, fines, or additional remediation requirements.

- *Risk of Contaminant Leaching*: Depending on the site conditions and hydrological factors, there is a risk of contaminants leaching from the treated soil and potentially impacting groundwater or surface water. Careful site characterization, monitoring, and appropriate containment measures are necessary to prevent or minimize the risk of contaminant migration.

- *Residual Contaminant Levels*: Land farming may not achieve complete remediation, especially for soils with high initial contamination levels or certain types of contaminants. In some cases, residual levels of contaminants may remain in the soil after land farming, necessitating additional remediation or long-term monitoring.

4.2.2. Biopile: Soil biopiles, also known as biocells, is a biodegradation technique used for the remediation of excavated soil contaminated with petroleum contents. This technology involves the accumulation of contaminated soil into piles and the stimulation of microbial activity either aerobically or by the addition of nutrients, minerals or moisture. A typical height of biopiles is between three and ten feet (Jørgensen et al., 2000).

Biopiles are in a way similar to landfarms due to the fact that this technology also uses oxygen as a way to stimulate bacterial growth. However, while tilling or plowing aerates land farms, biopiles are aerated by forcing air to move by injection through perforated piping placed throughout the pile (EPA, 2003). The contaminated soil is piled up to a depth of a few feet and then the piping, which provides aeration, is laid down. The next load of contaminated soil is then added. This process continues until the desired pile height is achieved. The soil is usually mixed with a bulking agent (straws) to improve aeration and therefore enhance the growth of the microbial population.

Since air is also injected into the soil there is the possibility for evaporation or volatization of contaminants. To counter this problem, the system also includes the monitoring and containment of soil vapors.

Biopile-mediated bioremediation involves above-ground piling of excavated polluted soil, followed by nutrient amendment, and sometimes aeration to enhance bioremediation by basically increasing microbial activities. The components of this technique are: aeration, irrigation, nutrient and leachate collection systems, and a treatment bed. The use of this particular ex situ technique is increasingly being considered due to its constructive features including cost effectiveness, which enables effective biodegradation on the condition that nutrient, temperature and aeration are adequately controlled. Biopile treatment has been applied to the treatment of non-chlorinated VOCs and fuel-contaminated soil. Chlorinated VOCs, semi-volatile organic compounds (SVOCs), and pesticides can also be treated, but process effectiveness varies.

Application of Biopile

» *Petroleum Hydrocarbon Contamination*: Biopiling is commonly used for the treatment of soil contaminated with petroleum hydrocarbons, such as crude oil, gasoline, diesel, or other similar substances. The contaminated soil is excavated and placed in piles, typically on an impermeable liner or a biopile pad. Amendments, such as bulking agents (e.g., wood chips), fertilizers, and oxygen sources, are added to the piles to optimize microbial growth and hydrocarbon degradation.

- *Refinery and Industrial Sites*: Biopiling is frequently applied at refinery and industrial sites where soil contamination occurs due to spills, leaks, or improper handling of petroleum products. The technique allows for the efficient treatment of large volumes of contaminated soil in a controlled manner. Biopiles can be constructed on-site, reducing the need for off-site transportation and disposal of contaminated soil.

- *Polycyclic Aromatic Hydrocarbons (PAHs)*: Biopiling can be effective for the treatment of soils contaminated with PAHs, which are persistent organic pollutants often associated with industrial activities, coal tar, or creosote. Specific amendments, such as compost or specialized microbial cultures, may be added to the biopiles to promote the growth of microorganisms capable of degrading PAHs.

- *Landfill Remediation*: Biopiling can be utilized as part of the remediation strategy for landfill sites contaminated with organic materials. Excavated soil or waste material from the landfill is placed in biopiles and treated with amendments to facilitate the breakdown of organic contaminants. The technique helps to accelerate the natural degradation processes and reduce the overall environmental impact of the landfill. (Gandhi et al., 2022)

- *Agricultural and Livestock Operations*: Biopiling can also be applied in agricultural or livestock operations to manage and treat organic waste materials, such as manure or compost. By constructing biopiles with appropriate amendments, the organic matter can be efficiently decomposed, reducing odor, pathogens, and potential environmental impacts.

Advantage of Biopile

- *Enhanced Contaminant Degradation*: Biopiling creates an optimized environment for microbial activity, promoting the degradation of contaminants. The addition of amendments, such as bulking agents, nutrients, and oxygen sources, enhances the growth and activity of contaminant-degrading microorganisms. This leads to accelerated and efficient degradation of organic pollutants, including petroleum hydrocarbons, PAHs, and other organic compounds.

- *Cost-effectiveness*: Biopiling is often a cost-effective remediation option, particularly for moderately contaminated sites. It generally requires lower capital and operational costs compared to more intensive remediation technologies. The simplicity of the method, reduced reliance on expensive equipment or chemical additives, and minimal site disruption contribute to its cost-effectiveness.

- *Minimal Site Disruption*: Biopiling can be conducted on-site, minimizing the need for soil excavation and off-site disposal. This reduces site disruption, transportation costs, and potential risks associated with soil transportation. The treatment is performed in designated piles or mounds, allowing for treatment to occur without disturbing the surrounding environment or ecosystems.
- *Flexibility and Scalability*: Biopiling is a flexible and scalable technique that can be adapted to different site conditions and contaminant types. It can be applied to both small- and large-scale projects, and the design and configuration of the biopiles can be adjusted accordingly. The method is applicable to various types of organic contaminants and can be customized based on specific treatment goals and site requirements.
- *Environmental Compatibility*: Biopiling is an environmentally friendly remediation method as it relies on natural microbial processes to degrade contaminants. It avoids the use of harsh chemicals or invasive techniques, reducing the generation of waste or by-products. It is a sustainable approach that utilizes the existing microbial populations in the soil to carry out the biodegradation process.
- *Reuse of Treated Soil:* After the completion of biopiling treatment, the treated soil can often be reused on-site or in other applications. Depending on the contaminant concentrations and regulatory requirements, the soil may meet the necessary criteria for reuse. This can help minimize waste generation and reduce the need for additional soil resources.
- *Regulatory Acceptance:* Biopiling is a well-established and accepted remediation technique for organic contamination. It has a proven track record of success in various remediation projects, and its effectiveness and reliability are recognized by regulatory agencies. This can streamline the permitting and regulatory processes associated with remediation projects.

Limitation of Biopile

While biopiling is a useful bioremediation technique, it also has certain limitations. Here are some of the limitations of biopiling:

- *Contaminant Limitations*: Biopiling is most effective for the treatment of organic contaminants, such as petroleum hydrocarbons or polycyclic aromatic hydrocarbons (PAHs). It may not be suitable for the remediation of soils contaminated with heavy metals, radioactive materials, or certain persistent organic pollutants. Different contaminants require specific remediation approaches, and biopiling may not be the optimal solution for all types of soil contamination.

» *Depth Limitations*: Biopiling is typically effective for shallow contaminant plumes or surface contamination. The treatment efficiency decreases with depth, making it less suitable for deeply buried contaminants or contaminated groundwater. In such cases, alternative remediation techniques like pump-and-treat or in-situ bioremediation may be more appropriate.

» *Heterogeneous Soils*: The effectiveness of biopiling can be reduced in soils with significant heterogeneity in terms of texture, composition, or moisture content. Variations in soil properties can lead to non-uniform distribution of amendments and microbial populations, affecting the overall treatment efficiency. Site characterization and proper design are crucial to address these challenges.

» *Nutrient Imbalance*: Biopiling requires the addition of amendments to provide necessary nutrients for microbial growth and degradation. However, achieving the optimal nutrient balance can be challenging. Excessive or insufficient nutrient amendments can impact microbial activity and treatment efficiency. Careful consideration of the specific contaminant, site conditions, and monitoring of nutrient levels is necessary for successful biopile implementation.

» *Climate Constraints*: Climate conditions, such as extreme temperatures, rainfall, or arid environments, can affect the performance of biopiling. Microbial activity and degradation rates are influenced by temperature, moisture content, and oxygen availability. Extreme conditions can limit microbial growth and reduce the overall treatment effectiveness. Adaptations, such as insulation or covers, may be required to mitigate the impacts of climate constraints.

» *Long Treatment Time*: Biopiling is a time-consuming process that can take several months to years to achieve significant contaminant reduction. The length of treatment time depends on factors such as the initial contaminant concentration, microbial activity, and site-specific conditions. In situations where a rapid cleanup is necessary, alternative remediation methods may be more suitable.

» *Monitoring and Maintenance*: Biopiling requires regular monitoring and maintenance throughout the remediation process. Parameters such as temperature, moisture content, oxygen levels, and contaminant concentrations need to be monitored and adjusted as necessary. Maintenance activities, such as turning the piles or adding amendments, may be required to optimize microbial activity and treatment efficiency.

4.2.3. Bioreactor

» As the name suggests, a bioreactor is a vessel where a series of biological reactions transform basic materials into one or more products.

» Batch, fed-batch, sequencing batch, continuous, and multistage are some of the several bioreactor operation modes.

» The choice of operating mode is mostly influenced by capital expenditure and market economy.

» In order to provide ideal growth circumstances, conditions in a bioreactor imitate and preserve the natural habitat of cells. (Robles-González et al., 2008).

» When compared to alternative ex situ bioremediation methods, using a bioreactor to treat contaminated soil offers a number of benefits. Polluted samples can be fed into a bioreactor as slurry or dry matter.

» One of the main benefits of bioreactor-based bioremediation is the excellent control of bioprocess parameters (temperature, pH, agitation and aeration rates, substrate and inoculum concentrations).

» It is implied that biological processes within a bioreactor can be improved to efficiently shorten the bioremediation period by controlling and manipulating process parameters.Importantly, controlled bioaugmentation, nutrient addition, increased pollutant bioavailability, and mass transfer (contact between pollutant and microbes),which are among the limiting factors of bioremediation process can effectively be established in a bioreactor thus making bioreactor-based bioremediation more efficient.

- Further, it can be used to treat soil or water polluted with volatile organic compounds (VOCs) including benzene, toluene, ethylbenzene and xylenes.
- The applications of different bioreactors for bioremediation process have resulted in removal of wide range of pollutants.
- A bioreactor is a vessel, wich is used to carry out one or more biochemical reactions to convert raw materials to productos through the action of biocatalyst, enzyme microorganisms, cells of animals or plants. The raw material could be an organic compound as sugar or an inorganic chemical as CO2.
- Bioreactors are very used in many of the biotechnology processes, we have different types of bioreactors everyone with their own advantage.

Application of Bioreactor in bioremediation

In ex-situ bioremediation, contaminated media such as soil, sediment, or groundwater is removed from the site and treated in a different location using various techniques, including the use of bioreactors. Here are some applications of bioreactors in ex-situ bioremediation:

- *Soil Bioremediation*: Contaminated soil can be excavated and placed in an ex-situ bioreactor system for treatment. The bioreactor provides a controlled environment where optimal conditions for microbial growth and contaminant degradation can be maintained. The soil is typically mixed or agitated periodically to promote contact between the microorganisms and contaminants. The bioreactor system can be designed as a pile, container, or treatment cell.
- *Slurry Bioremediation*: In slurry bioremediation, contaminated soil or sediment is mixed with water to create a slurry, which is then treated in an ex-situ bioreactor. The bioreactor provides suitable conditions for microbial activity, allowing for the biodegradation of contaminants. The slurry can be continuously mixed or agitated to ensure uniform contact between the contaminants and microorganisms. This technique is particularly useful for treating fine-grained or highly contaminated soils.
- *Pump-and-Treat Systems*: In pump-and-treat systems, contaminated groundwater is extracted from the site and treated in an ex-situ bioreactor. The bioreactor provides a controlled environment where specific microbial communities can degrade the contaminants present in the groundwater. The

system can be designed as a tank or column system, allowing for controlled flow rates, aeration, and nutrient supplementation to enhance microbial activity.

» *Bioreactors for Contaminated Water Treatment*: Bioreactors can be used to treat contaminated surface water or wastewater in ex-situ bioremediation applications. The water is directed into a bioreactor system where specific microbial communities degrade the contaminants present. The bioreactor environment can be optimized to target the specific contaminants of concern and achieve desired treatment outcomes. This application is commonly used for the removal of organic pollutants, nutrients, or specific target compounds from water sources.

» *Landfarming Bioreactors*: Landfarming, which involves the treatment of contaminated soil, can be enhanced by the use of bioreactors. The contaminated soil is spread out in a treatment cell or pile and subjected to controlled conditions within the bioreactor system. This can include aeration, moisture control, and nutrient supplementation to promote microbial degradation of contaminants.

Advantages of Bioreactor in bioremediation

Bioreactors offer several advantages when applied to ex-situ bioremediation processes (Tekere et al., 2019). Here are some of the key advantages of using bioreactors in ex-situ bioremediation:

» *Enhanced Contaminant Degradation*: Bioreactors provide a controlled environment that optimizes conditions for microbial growth and contaminant degradation. By adjusting factors such as temperature, pH, oxygen levels, and nutrient availability, bioreactors enhance the activity of contaminant-degrading microorganisms. This leads to improved degradation rates and more effective removal of pollutants compared to natural attenuation or traditional treatment approaches.

» *Increased Treatment Efficiency*: Bioreactors allow for more efficient and rapid treatment of contaminated media. The controlled conditions in bioreactors promote the growth of specific contaminant-degrading microorganisms, ensuring that they have the necessary resources to thrive and degrade the contaminants. This results in accelerated remediation timelines, allowing for quicker site cleanup and restoration.

» *Flexibility in Treatment Design*: Bioreactors offer flexibility in the design and configuration of treatment systems. The size, shape, and operational parameters of bioreactors can be adjusted to meet the specific requirements of the remediation project. This flexibility allows for the treatment of diverse contaminants, different types of media, and varying site conditions.

- *Contaminant-Specific Approach:* Bioreactors can be tailored to target specific contaminants and optimize their degradation. The design and operation of bioreactors can be optimized to address the unique characteristics of the contaminants, such as their chemical composition, concentration, and persistence. This specificity allows for efficient degradation of the target contaminants while minimizing the impact on non-targeted compounds.
- *Reduced Site Disruption*: Ex-situ bioremediation using bioreactors minimizes the disruption to the site. Instead of conducting treatment directly in the field, contaminated media can be transported to a designated treatment facility, reducing potential environmental risks associated with on-site treatment. This approach can be particularly advantageous for sites with limited access, sensitive ecosystems, or when in-situ remediation is not feasible.
- *Monitoring and Control*: Bioreactors allow for real-time monitoring and control of critical parameters such as temperature, pH, dissolved oxygen, and nutrient levels. This enables proactive management of the remediation process and the ability to optimize conditions for microbial activity. Continuous monitoring ensures that the bioremediation system is operating effectively and allows for prompt intervention if any issues arise.
- *Compliance with Regulatory Requirements*: Bioreactors provide a controlled treatment environment that can help meet regulatory requirements for the remediation of contaminated sites. By offering enhanced treatment efficiency and predictable outcomes, bioreactors can help ensure that remediation goals are achieved within the specified timeframes and meet the required cleanup standards.

Limitation of Bioreactor in bioremediation

While bioreactors are a valuable tool in ex-situ bioremediation, there are certain limitations and challenges associated with their use (Admassu et al., 1996). Some of the key limitations of bioreactors in ex-situ bioremediation include:

- *Cost*: Implementing and operating bioreactor systems for ex-situ bioremediation can be expensive. The costs include the design and construction of the bioreactor, installation and maintenance of monitoring equipment, energy requirements, and ongoing operation and management expenses. These costs can vary depending on the scale of the remediation project and the complexity of the contaminants being treated.

» *Space Requirements*: Bioreactors require adequate space for their installation and operation. Depending on the volume of contaminated media being treated, bioreactor systems can occupy significant areas. This may pose challenges in cases where there is limited available space or where land use restrictions are in place.

» *Site Accessibility and Logistics*: Transporting contaminated media from the site to the location of the bioreactor system can present logistical challenges. The excavation, handling, and transportation of large volumes of contaminated media can be time-consuming, expensive, and may require special permits or approvals. Site accessibility and proximity to the treatment facility also need to be considered to ensure the feasibility of ex-situ bioremediation using bioreactors.

» *Contaminant Heterogeneity*: The effectiveness of bioreactors in ex-situ bioremediation can be affected by the heterogeneity of contaminants within the media. Variations in contaminant distribution can result in uneven treatment and may require additional steps, such as pre-treatment or segregation of highly contaminated areas, to ensure effective remediation.

» *Microbial Adaptation and Nutrient Limitations*: Bioreactors rely on microbial activity to degrade contaminants. However, microbial populations may need time to adapt and establish themselves within the bioreactor system. Additionally, the availability of nutrients, such as carbon, nitrogen, and phosphorus, can affect microbial growth and activity. Ensuring adequate nutrient supply and maintaining optimal conditions for microbial activity can be challenging in some cases.

» *Technology Limitations:* The selection and design of bioreactors should be based on the specific characteristics of the contaminants and media being treated. Certain contaminants or complex mixtures of contaminants may require specialized bioreactor configurations or additional treatment steps to achieve effective remediation. It is essential to consider the compatibility of the bioreactor technology with the specific contaminants and site conditions to ensure successful treatment.

- *Long-Term Monitoring and Maintenance*: Ex-situ bioremediation using bioreactors requires ongoing monitoring and maintenance to ensure the effectiveness of the treatment. Regular monitoring of key parameters such as temperature, pH, nutrient levels, and contaminant concentrations is necessary to optimize performance and address any operational issues. Long-term maintenance and periodic system evaluation are also essential to ensure continued performance over time.

Disadvantages

- The need for shaft seals and bearings.
- Size limitation by motor size, shaft length and weight.

4.2.4. Bio-comosting

- Compost is organic matter that has been broken down into simpler organic or inorganic matter in a process called composting (Rosa et al., 2023).
- This process recycles various organic materials otherwise regarded as waste products and produces a soil conditioner.
- The term bio-compost means plant matter that has been decomposed and recycled as a fertilizer or manure.
- Bio-compost is considered as a key ingredient in organic farming.
- It is very rich in nutrients.
- The process bio-composting is done by simply piling up wastes in the field or any outdoor place and then leave it undisturbed for a year or more.
- Bio-compost in the ecosystems is very useful for control of soil erosion, wetland construction, and as landfill cover.
- Modern day bio-composting process has many steps like monitoring of the composting.
- It is usually done by shredding the plant matter, adding of sufficient water to maintain the proper moisture level and then regularly turning the mixture to provide better aeration.
- Addition of worms and fungi helps in the process of decomposition.

- They break up the complex compounds into simpler ones and during the process lots of heat, carbon dioxide and ammonium is produced.
- This ammonium is again utilized by the microbes which are made available to the plants as nitrites and nitrates.
- Thus, bio-composting is an easy and simple process for the decomposition of organic wastes which in turn can be used as manure or fertilizer. The process is also very cost effective.

The Principles of the Composting Process

The composting process can transform large quantities of organic material into compost in a relatively short time by properly managing moisture, air and nutrients. During composting, the microorganisms consume oxygen and feed on organic matter (Stentiford et al., 1996). Active composting generates a considerable amount of heat and large quantities of carbon dioxide (CO_2) and water vapour are released into the air. The CO_2 and water losses can amount to half the weight of the initial organic materials, so composting reduces both the volume and mass of the raw materials while transforming them into a beneficial humus-like material.

$$\text{Organic compounds} + O_2 \xrightarrow{\text{Microbial metabolism}} \text{Stablized organic waste material} + CO_2 + 2H_2O + \text{Energy}$$

Types of Composting

Passive Composting

- Composting is a dynamic process
- Depending on the process used it will occur quickly or slowly
- A pile of organic waste will eventually decompose, but slowly. (Decomposition Happens)
- "Passive composting" requires very little maintenance.

Active Composting

- Fast or "active" composting can be completed by creating ideal conditions to encourage the natural decomposition
- Active Composters pay attention to:

- Aeration
- Moisture
- Carbon to nitrogen (C:N) ratio

Why Composting

» Waste Minimization

» Nutrient and Water Storage

» Improved Soil Structure

» Erosion Prevention

» Disease Resistance

» Bio-composting not only solves the disposal problems of effluent but also helps in saving the cost on chemical fertilizers.

Different Stages during Composting

MESOPHILIC PHASE (20-40°C)
Bacteria and thermotolerant fungi

↓

INITIAL THERMOPHILIC PHASE (40-60°C)
Thermophilic bacteria (e.g., Bacillus), actinomycetes (e.g., Streptomyces) and fungi *(e.g., Aspergillus)*

↓

THERMOPHILIC PHASE (60-80°C)
Thermophilic spore- forming bacteria (e.g., *Bacillus* spp.), Sulfur and hydrogen-oxidizing autotrophic, bacteria (e.g., Hydrogenobacter spp.), Heterotrophic, aerobic, non-spore-forming, Bacteria (e.g., Themus spp.)

↓

COOLING & MATURATION PHASES
Bacteria involved in nutrient cycling, Mesophilic/thermotolerant actinomycetes and fungi

The Science of Composting

Mesophilic Phase 1 (10-40 ^{0}C)

- Lasts only a few days
- Explosive growth of bacteria and fungi
- Rapid breakdown of soluble sugar and starches

Thermophilic Phase (>40 0 C)

- Can last from several days to several months depending on size of system
- Mixed population of heat loving organisms
- High heat helps breakdown of proteins, fats, "tough" plant material like cellulose
- High temperature (>55 °C) kill weeds and pathogen harmful to humans
- Higher temperature (>60°C) kill organism needed for decomposition

Mesophilic Phase 2 (10-40 0 C) "Curing Phase"

- Can last several months
- Bacteria, fungi, actinomycetes (mix between bacteria and fungus, give "earthy" smell) predominate. Invertebrates active.
- Supply of organic material has decreased. Remaining organic material is slowly broken down.
- Additional chemical reactions take place to make remaining organic material more stable

Benefits of bio compost Soil

- Improves its physical structure
- Enriches soil with micro-organisms
- Microbial activity in worm castings is 10 to 20 times higher than in the soil and organic matter that the worm ingests
- Attracts deep-burrowing earthworms already present in the soil
- Improves water holding capacity

- Plant Growth:Enhances germination, plant growth and crop yield
- Improves root growth and structure
- Enriches soil with micro-organisms

Economic

- Biowastes conversion reduces waste flow to landfills
- Elimination of bio-wastes from the waste stream reduces contamination of other recyclables collected in a single bin.
- Creates low-skill jobs at rural level
- Low capital investment and relatively simple technologies make vermicomposting practical for less-developed agricultural regions.

Environmental

- Helps to close the "metabolic gap" through recycling waste on-site
- Large systems often use temperature control and mechanized harvesting, however other equipment is relatively simple and does not wear out quickly
- Production reduces greenhouse gas emissions such as methane and nitric oxide.

Uses of bio compost

- Compost is generally recommended as an additive to soil, or other matrices such as coir and peat, as a tilth improver, supplying humus and nutrients.
- It provides a rich growing medium, or a porous, absorbent material that holds moisture and soluble minerals, providing the support and nutrients in which plants can flourish, although it is rarely used alone, being primarily mixed with soil, sand, grit, bark chips, vermiculite, and clay granules to produce loam (Vieira et al., 2019).
- Compost can be tilled directly into the soil or growing medium to boost the level of organic matter and the overall fertility of the soil. Compost that is ready to be used as an additive is dark brown or even black with an earthy smell.
- Composting can destroy pathogens or unwanted seeds. Unwanted living plants (or weeds) can be discouraged by covering with mulch/compost.

Advantages of Composting

- » Proceeds relatively quickly under ideal conditions.
- » Weed seeds and pathogens are killed during the process.
- » Proceeds easily in cold weather on a large scale.

Disadvantages of Composting

- » Requires attention and is labor intensive.
- » May require some stock-piling until sufficient materials are available for a "batch."
- » Heating can lead to considerable nitrogen loss.
- » Heat can kill off many beneficial microbes.

4.2.5. Constructed wetland

- » Constructed wetland treatment systems efficiently remove several kinds of pollutants from wastewater, drinking water, and industrial effluents within the controlled environment and are considered as environment-friendly (Parde et al., 2021).
- » Removal of effluents, metal and waste are performed through a variety of processes i.e., by increasing sorption, hydrolysis, filtration and oxidation, precipitation, binding with iron oxide, microbial activity and uptake by plants.
- » In response to removing pollution, constructed wetlands differ in their processes, cost effectiveness and their design and also enhance the waste removal performance by using different substrates i.e., Phosphorus (P).
- » In the recent years, scientists have tried to remediate or clean up wastewater through phytoremediation and bioremediation.
- » Constructed wetland treatment systems use microbes and rooted plants to remove contaminants from soil or wastewater.
- » It takes advantage of natural wetland processes (biological, physical and chemical processes to remove contaminants but the efficiency of all the processes (Chemical, physical and Biological) differ with water residence time.

- In fluent waters flow through and beneath the gravel surface of a gravel-based wetland.
- The wetland (with plants) is a coupled anaerobic-aerobic system.
- The anaerobic cell uses plants with natural microbes to degrade the contaminant.
- The aerobic (reciprocating) cell further improves water quality through continued exposure to the plants and the movement of water between cell compartments.
- Most commonly been used in wastewater treatment for organic matter; nutrients (nitrogen and phosphorus), suspended sediments & for control of trace metals and other toxic materials.
- Has been used to treat acid mine drainage - typically acidic with high metal concentrations (from metal or coal mining activity)

Limitations

- Unsure of long-term effectiveness of constructed wetlands - may decrease in contaminant removal rates over time.
- Possibly very high cost to construct artificial wetland - varies considerably for each site.
- Temperature and fluctuations in flow affect wetland function - inconsistent contaminant removal rates.
- Colder conditions slow the rate of degradation.
- A heavy inflow of water can overload the removal mechanisms in a wetland
- A dry spell can damage plants - severely limiting wetland function.
- Wetland treatment is a long-term technology intended to operate continuously for years. Many studies and successful applications have been done in the treatment of contaminated soil and water. The activity of microbial can be affected by the following factors: temperature, oxygen, pH and nutrients. For successful biodegradation, microorganisms should develop a catabolic activity by the following processes: new metabolic capabilities development by changes of genetic, induction of specific enzymes and eclectic enrichment of microorganisms that are capable to convert the pollutants.

Chapter - 5

Concept and Mechanism of Phytoremediation

5.1. Introduction

Phytoremediation is a word made from the Greek prefix "Phyto" meaning plant, and the Latin suffix "remedium" meaning clean or restore. The creative use of plants and related soil microbes to extract, sequester, detoxify, or immobilize elemental and organic chemical contaminants is known as phytoremediation (Shen et al., 2022). This fascinating and recently developed area of plant biology, environmental biology, and biotechnology offers enormous potential for cleaning up the contaminated soil, sediment, and water. Phytoremediation is gaining traction as a sustainable, eco-friendly "Green Clean" technology as an alternative to environmentally harmful physical remediation techniques like excavation and reburial, incineration, or capping of toxic sediments because it is founded on the idea of using nature to cleanse nature. Phytoremediation is an in-situ treatment that preserves top soil, reduces environmental disturbance, and enhances ecological habitat since it is powered by solar energy. Metals, pesticides, solvents, explosives, crude oil, poly aromatic hydrocarbons, and landfill leachates can all be cleaned up via phytoremediation. To deal with the accumulated contaminants in the environment, plants have developed a wide variety of genetic adaptations. Plants are special organisms with exceptional metabolic and absorption capacities as well as transport systems that can selectively absorb nutrients or pollutants from the soil, water, or growth matrix. In order to eliminate pollutants or aid in their immobilization (binding/containment) or phytodegradation (detoxification), phytoremediation entails cultivating plants in contaminated soil, water, and sediments for the necessary growth period. The plants can be subsequently harvested, processed and disposed. Phytoremediation has been successfully tested in many locations and successful outcomes have been achieved (Gavrilescu, 2022).

DOI: 10.1201/9781003787549-5

Large volumes of chemicals, garbage, and harmful gasses are released into the environment by anthropogenic activities, changing the physical, chemical, and biological characteristics of ecosystems all over the world. As part of their metabolism, biological agents use these pollutants, converting them into less harmful or innocuous byproducts. It represents the innate capacity of some plants known as hyperaccumulators to remove various pollutants from soil, water, or air, such as heavy metals, insecticides, and polyaromatic hydrocarbons. The habitats, nutrient supply, kind of contaminants, and environmental circumstances all influence which process—biotransformation, phytodegradation, phytostimulation, phytovolatilization, etc is selected for cleanup. Phytoremediation is an efficient and less expensive method of managing waste since it uses living organisms and guarantees the full breakdown of different toxicants without generating hazardous byproducts. However, phytoremediation is only effective in the plant's root zone. Additionally, this technology's use is restricted in areas where pollutant concentrations are harmful to plants (Wei et al., 2021). Technologies for phytoremediation are available for a range of pollutants and settings. These include a variety of procedures such in situ stabilization, degradation, and contaminant removal (i.e., volatilization or extraction).

5.2. Factors of Phytoremediation

In addition to creating biomass, plants can absorb, break down, change, and sequester pollutants. The parameters of the site and, most importantly, the type of pollutant determine the length of phytoremediation as well as the kind and quantity of species needed at a specific location (Magdziak et al., 2015). When using phytoremediation at a location, the following aspects are crucial to consider:

- Plant Selection
- Contaminant Characteristics
- Site Conditions
- Planting and Maintenance
- Duration and Monitoring
- Regulatory and Stakeholder Considerations
- Integration with Other Remediation Techniques
- Economic and Cost Considerations

Plant selection: The choice of plant species is crucial. Certain plants have a higher tolerance for specific contaminants and exhibit better remediation capabilities. Factors to consider include the ability of plants to uptake and tolerate contaminants, their biomass production, growth rate, root depth, and their potential for pollutant accumulation or degradation.

Contaminant Characteristics: Understanding the nature of the contaminants present is essential. Factors such as the type, concentration, mobility, and availability of contaminants in the environment influence the choice of plant species and the phytoremediation mechanism to be employed.

Site Conditions: Site-specific factors play a significant role in phytoremediation. These include soil characteristics (e.g., pH, texture, nutrient content), hydrological conditions (e.g., water availability, drainage), climatic factors (e.g., temperature, rainfall), and the presence of other plants or vegetation. Proper site characterization is necessary to select appropriate plant species and ensure their successful growth and pollutant uptake.

Planting and Maintenance: Proper planting techniques, including seed selection, transplanting, and establishment of plants, are crucial for a successful phytoremediation project. Adequate care and maintenance, such as irrigation, fertilization, weed control, and pest management, are important to ensure the health and growth of the plants.

Duration and Monitoring: Phytoremediation is a time-dependent process. The duration required for significant remediation varies depending on the contaminants, plant species, and site conditions. Regular monitoring is necessary to assess the progress of the remediation, measure contaminant concentrations, and adjust the management practices if needed.

Regulatory and Stakeholder Considerations: Compliance with applicable regulations, permits, and guidelines is essential when implementing a phytoremediation project. Involving stakeholders and obtaining their support and understanding can contribute to the success and acceptance of the remediation efforts.

Integration with Other Remediation Techniques: Phytoremediation can be combined with other remediation methods to enhance its effectiveness. Integration with techniques such as soil amendments, bioaugmentation, or physical remediation methods may be necessary for certain contaminants or site conditions.

Economic and Cost Considerations: Assessing the economic feasibility and cost-effectiveness of phytoremediation is crucial. It is important to consider factors such as the availability and cost of plant materials, labor, monitoring equipment, and long-term maintenance requirements.

By considering these factors and tailoring phytoremediation strategies to site-specific conditions, it is possible to maximize the efficiency and success of the remediation process.

Table1. Basic mechanisms and principles that are part of phytoremediation technologies

Mechanisms Principles	**Mechanisms Principles**
Heavy metals and inorganic contaminants	
Uptake and transport	- Plants and metals interact in the root environment; - The contaminant has to be in contact with the roots in order to be uptaken; - The same kind of transporters that are used for the macronutrients and micronutrients entrance are also used by plants for the heavy metals uptake; - Metals are stored in the subcellular compartments like vacuoles and lignocellulosic material (cell wall); - Research is being developed to discover other mechanisms that may be active to direct the metals in cells.
Accumulation and sequestration	- Analytical techniques (X-ray emission - scanning electron microscopy, microanalysis) can provide information on speciation and localization of metals in plants tissue; - It is important to understand the molecular bases of the capacity of some plants to hyperaccumulate and store the metals.
Identification of genes and proteins involved in tolerance and accumulation	- Genomics and proteomics approaches are being used to study the molecular and biochemical mechanisms of the process of hyperaccumulation.
Genetic bases of tolerance	- Genetic studies are being taken in order to determine the genetic bases of tolerance, accumulation; Gene identification studies using genomic and proteomic approaches.
New contaminants	- Most common contaminants include radionuclides, heavy metals and nutrients; - New contaminants that are subject of research: arsenic, mercury.

Organic contaminants	
Mechanisms of genetic controls – candidate genes	- Gene identification studies using genomic approaches; - In the xenobiotics metabolism, specific candidate genes can code for enzymes that are part of the process.
Analysis and identification of enzymes and proteins	- In the metabolism of xenobiotics, the identification of new enzymes is determined due to proteomic approaches.
Transgenic approaches	- Transgenic approaches that can modify/improve the enzyme that is responsible for metabolic modification and degradation of xenobiotic molecule.
In vitro studies for implementation	- Bioassessment studies that use cell cultures are being taken before practical application when it comes to more complex contaminants.

5.3. Phytoremediation Mechanisms of Heavy Metals

A living plant can be thought of as a solar-powered pump since it can collect and concentrate specific elements from a contaminated environment. After harvesting contaminated plants, the following techniques may be used: drying, ashing, or composting. The success of phytoremediation is dependent on considering as many aspects as possible while implementing phytoremediation initiatives.

Plant biochemical mechanisms in the case of phytoremediation of heavy metals are (Kaur et al., 2022)

1. *Adsorption:* root surface is the key factor because it absorbs the elemental nutrients, binds pollutants and nutrients and favors the interaction between plant roots and soil microbes with the aim of increasing metal bioavailability;
2. *Accumulation and transport:* role of transporter proteins
   - proteins and peptides that increase metal binding in plants can improve metal tolerance or accumulation; chelating agents (natural and synthetic chelators)
   - chelating agents added to the soil help increase the metal bioavailability, uptake and translocation of heavy metals;
3. *Translocation:* root cells uptake metal ions and transport them to the shoots, process in which the membrane transport systems have a major role;
4. *Detoxification:* hyperaccumulators possess a great characteristic of being very efficient in detoxification and sequestration without having phytotoxic effects due to the adsorption of huge amounts of heavy metals;
   - vacuolar compartmentalization

- » vacuoles are the main storage place of heavy metals in the cells of a plant and there is a vacuolar compartmentalization to control the distribution and concentration of metal ions in order to restrict other parts of the cell to have access to the contaminants;
- » volatilization – metal ions are converted into volatile state;

5. *Hyperaccumulation:* the metal ion is concentrated to > 0.1 - 1 % of the dry weight of the plant.

5.4. Different mechanisms of phytoremediation

5.4.1. Phytoextraction/Phytoaccumulation:

Phytoextraction uses the ability of plants to accumulate contaminants in the aboveground, harvestable biomass. This process involves repeated harvesting of the biomass in order to lower the concentration of contaminants in the soil. Phytoextraction is either a continuous process (using metal hyperaccumulating plants, or fast growing plants), or an induced process (using chemicals to increase the bioavailability of metals in the soil). Continuous phytoextraction is based on the ability of certain plants to gradually accumulate contaminants (mainly metals) into their biomass. Generally, phytoextraction process involves the five major steps: pollutant mobilization in rhizosphere, pollutant uptake by plant roots, translocation toward aerial plant parts, and pollutant sequestration in plant tissues. Ideally a hyperaccumulator plant should possess the following characteristics: (i) high growth rate, (ii) more aboveground biomass, (iii) tolerance to a high concentrations of inorganic pollutants, (iv) widely distributed and highly branched root system, (v) adaptation to the local environmental conditions, (vi) translocation of the accumulated pollutants from roots to shoots, (vii) resistant to pathogens and pests, (viii) easy cultivation and harvest, and (ix) more accumulation of the target pollutants (Robinson et al., 2015).

Certain plants can hyperaccumulate metals without any toxic effects. These plants are adapted to naturally occurring, metalliferous soils. More than 400 plant species can hyperaccumulate various metals. However, most plants can only hyperaccumulate one specific metal.

The phytoextraction mechanism involves several steps:

Step 1: *Uptake*: The first step is the uptake of contaminants from the soil or water by the plant's roots. This uptake can occur through various mechanisms, including passive diffusion, active transport, or ion exchange. The availability and mobility of the contaminants in the soil play a significant role in their uptake.

Step 2: *Translocation*: Once the contaminants are absorbed by the roots, they are transported through the plant's vascular system to the above-ground tissues. This translocation occurs through the xylem, which is responsible for transporting water and dissolved substances from the roots to the shoots. Contaminants are carried along with water as it moves through the plant.

Step 3:*Accumulation*: As the contaminants are translocated to the above-ground tissues, they can accumulate and be stored in various plant parts. This accumulation can occur in the leaves, stems, or other plant organs, depending on the specific contaminant and plant species. Some plants have specialized structures or compartments, such as trichomes or vacuoles, where they sequester the accumulated contaminants.

Step 4: *Harvesting and Disposal*: Once the plants have accumulated a significant amount of contaminants, they can be harvested and properly disposed of to remove the pollutants from the environment. Harvesting can involve cutting the above-ground biomass or uprooting the entire plant, depending on the specific objectives of the phytoremediation project.

There are some hyperaccumulator plants which can accumulate different metals upto 1000mg/L from the polluted sites. The common examples of hyperaccumulator plants are given below:

Uptake
(Uptake of contaminants from the soil or water)

↓

Translocation
(Transportation of the contaminant to the above-ground tissues)

↓

Accumulation
(Accumulated to the different part of plant tissue

↓

Harvesting, phytominig and Disposal
(plant can be harvested either for metal recovery or disposal)

Flow diagram showing phytoextraction steps and its disposal

Table 5.2: Some recorded levels of hyperaccumulator plant species with their metal concentration

Metal	**Plant species**	**mg/kg**	**Reference**
Ni	*Isatis pinnatiloba*	1441	Altinozlu et al., 2012
Ni	*Alyssum serpyllifolium*	10000	Prasad, 2005
Ni	*Alyssum pterocarpum*	13500	Li et al., 2003
As	*Corrigiola telephiifolia*	2110	Garcia-Salgado et al., 2012
Cu	*Eleocharis acicularis*	20200	Sakakibara et al., 2011
Pb	*Euphorbia cheiradenia*	1138	Chehregani and Malayeri, 2007
As	*Pteris biaurita*	~2000	Srivastava et al., 2006
Mn	*Schima superba*	62412.3	Yang et al., 2008
Ni	*Alyssum markgrafii*	19100	Bani et al., 2010
As	*Petris vittata*	8331	Shweta et al., 2011
Cs	*Amaranthus chlorostachys*	2146	Moogouei et al., 2011
Cd	*Thlaspi caerulescens*	14,000	Sachs, 1865
Se	*Astragalus bisulcatus*	10,000	Freeman et al., 2006
Tl	*Biscutella laevigate*	14,000	Anderson et al., 1999
U	*Brassica juncea*	5000	Huang et al., 1998
V	*Brassica juncea*	10,111	Elektorowicz et al., 2015
Zn	*Arabis paniculate*	20,800	Tang et al., 2009
Cu	Aeolanthus biformifolius	9000	Morrison et al., 1979
Co	*Haumaniastrum robertii*	10,200	Dodson et al., 2012
As	*Pteris vittata*	22,630	Wang et al., 2007
Li	Beta vulgaris	5500	Li et al., 2009

Hyperaccumulating plants can contain more than 1% of a metal in their dry biomass. For example, the hyperaccumulating plant *Berkheya coddii* was found to contain as much as 3.8% of Ni in the dry, aboveground biomass, when grown in contaminated soil. It is possible to extract metals from the harvested biomass in a process termed phytomining.

Phytomining is a method of extracting valuable metals or minerals from the soil using plants. Certain plant species, called hyperaccumulators, have the ability to absorb high concentrations of metals from the soil through their roots. These metals accumulate in the plant's tissues, and when the plants are harvested, the metals can be extracted from the biomass. Phytomining is an alternative to traditional mining methods and offers a more environmentally friendly approach, as it reduces the need for extensive excavation and reduces the release of harmful pollutants into the environment. However, phytomining is currently a niche technique and is not widely used on a commercial scale.

The underlying mechanism of hyperaccumulation of metals in plants is the overexpression of genes that regulate cell membrane transporters. These include the Cu-transporter (COPT1) and Zn-transporter (ZNT1). The main limitations on the use of hyperaccumulating plants in phytoextraction are slow growth and low biomass production. The effectiveness of phytoextraction is a function of a plant's biomass production and the content of contaminants in the harvested biomass. Therefore, fast-growing crops that accumulate metals have a great potential in phytoextraction. The use of crops in phytoextraction can be improved by manipulation of their associated soil microbes. Inoculation of plant growth promoting bacteria (PGPR) and Arbuscular Mycorrhizal fungi (AMF) can increase plant biomass. The AMF-plant symbiosis usually results in reduced accumulation of metals in the aboveground biomass of plants. Therefore, suppressing AMF activity, by using specific soil fungicides, has resulted in increased metal accumulation in plants. The role of AMF in regulating metal uptake by plants appears to vary depending on numerous factors, such as AMF populations, plant species, nutrient availability, and metal content in the soil. Also, this regulation of AMF is usually metal-specific; where the uptake of essential metals is generally increased, but the uptake of non-essential metals is inhibited.

However, exceptions have been found where AMF increases uptake of Ni, Pb, and As in plants. Induced phytoextraction involves the use of fast-growing crops and chemical manipulation of the soil. Low bioavailability of metals in the soil is a limiting factor in phytoextraction. The bioavailability of metals can be increased by the use of synthetic chelates such as ethylenediaminetetracetic acid (EDTA) or acidifying chemicals (e.g., NH_4SO_4). The use of synthetic chelates increases the absorption of metals to the root and the translocation of metals from the roots to the foliage. The timing of chelate application is critical, and should ideally take place at the peak of biomass production. The effectiveness of using EDTA was demonstrated by growing corn (*Zea mays*) in Pb-contaminated soil treated with 10 mmol kg^{-1} EDTA. This resulted in a high accumulation of Pb (1.6% of shoot dry weight), and facilitated the translocation of Pb from the roots to the foliage. Some drawbacks of using synthetic chelates in phytoremediation are the result of increased solubility of the metals within the soil. In turn this increases the risk of metal migration through the soil profile and into the groundwater. However, a possible solution is to treat contaminated soil ex situ in a confined site with an impervious surface. Also, periodic application of low doses of synthetic chelates reduces the risk of metal migration.

5.4.2. Phytostabilization

Phytostabilization aims to retain contaminants in the soil and prevent further dispersal. Phytostabilization may be characterized as restriction of a pollutant in the contaminated media in the course of assimilation and aggregation by roots, adsorption onto roots, or precipitation within the root region of plants and (2) the deployment of plants and plant roots to avoid contaminant movement through wind and water, draining, and dispersion of soil (Bolan et al., 2011).

Unfortunately, phytostabilization is not a permanent solution to contamination, because phytostabilization does not reduce the concentration of pollutants but reduces the contamination of nearby media/area. Contaminants can be stabilized in the roots or within the rhizosphere.

The phytostabilization mechanism involves the following steps:

Step1: *Uptake*: Similar to other phytoremediation mechanisms, phytostabilization begins with the uptake of contaminants from the soil by plant roots. However, in phytostabilization, the primary goal is not to accumulate the contaminants in the plant tissues but rather to prevent their movement and reduce their availability for uptake.

Step 2: *Binding and Complexation:* Once taken up by the roots, the contaminants may undergo various processes within the plant. Some plants have the ability to bind and complex the contaminants in their root zone. This can occur through the release of root exudates, which contain organic compounds capable of forming complexes with the contaminants, reducing their mobility.

Step 3: *Encapsulation*: In certain cases, plants can produce substances or compounds that encapsulate the contaminants in the soil. These substances may include polysaccharides, lignin, or other organic materials secreted by the roots. The encapsulation helps to immobilize the contaminants, reducing their bioavailability.

Step 4: *Rhizodegradation*: Phytostabilization can also involve the release of enzymes or compounds by plant roots, which can facilitate the degradation or transformation of contaminants in the rhizosphere (the soil zone influenced by the plant roots). These enzymatic activities can break down or convert contaminants into less toxic or less mobile forms.

Step 4: *Root Zone Modification*: Plant roots can physically modify the soil structure and properties, creating conditions that promote the immobilization or sequestration of contaminants. For example, roots can increase soil organic matter content, enhance soil aggregation, or change soil pH, which can influence contaminant mobility and availability. The mechanism for phytostabilization is shown in flow chart below.

Uptake
(Uptake of contaminants from the soil or water)

↓

Binding and Complexation
(Bind and complex the contaminants in their root zone)

↓

Encapsulation
(Plants can produce substances or compounds that encapsulate the contaminants in the soil)

↓

Rhizodegradation
(Plants can produce substances or compounds that encapsulate the contaminants in the soil)

↓

Root Zone Modification
(Plant roots can physically modify the soil structure and properties, creating conditions that promote the immobilization or sequestration of contaminants)

Flowchart showing the mechanism of phytostabilization

Some of the examples of phytostabilization in minimizing the industrial and polluted sites have been noted. Mine tailings have been stabilized using commercially available varieties of metal tolerant grasses such as *Agrostis tenuis* cv. Goginan and cv. Parys and *Festuca rubra* cv. Merlin. Metal tolerance of plants is generally increased by symbiotic, root-colonizing, arbuscular mycorrhizal fungi (AMF), through metal sequestration in the AMF hyphae. In addition, excretion of the glycoprotein glomulin by AMF hyphae can complex metals in the soil. Populations of AMF that are adapted to metal contaminated soils have great potential in phytostabilization.

Soil microbes can decrease toxic effects of contaminants in the soil. For example, exudates (peptides) from the bacterium *Pseudomonas putida* can decrease Cd toxicity in plants. Plants can also convert contaminants into less toxic forms, or decrease their bioavailability. Natural chelates released by the roots of certain plants can form complexes with metals in the rhizosphere. These include siderophores, organic acids and phenolics. Metals such as the toxic Cr (lll) can be converted to the much less toxic Cr(Vl) by enzymes found on the roots of wetland plants. In addition, plants, and their associated soil microbes, can release chemicals that act as biosurfactants in the soil that increase the uptake of contaminants. Contaminants can be stabilized in natural and constructed wetlands through a process called phytofiltration. This involves rhizofiltration where metals are precipitated within the rhizosphere. Metal-plaque forms typically on the roots of wetland plants through the release of oxygen via the aerenchyma of roots. Iron oxides precipitate along with other metals into the metal plaque. Metal plaque on roots acts as a reservoir for active iron (Fe2+), which in turn increases the tolerance of plants to other toxic metals.

5.4.3. Phytodegradation:

Phytodegradation involves the degradation of organic contaminants directly, through the release of enzymes from roots, or through metabolic activities within plant tissues. This process relies on the plants' ability to produce enzymes or harbor microorganisms that can metabolize or transform pollutants into less harmful substances (Newman and Reynolds, 2004). It is particularly effective for organic contaminants such as petroleum hydrocarbons, pesticides, and certain industrial chemicals.

The phytodegradation mechanism involves several steps:

Step 1: *Uptake*: Contaminants present in the soil or water are taken up by the roots of the plants. This uptake can occur through passive diffusion or active transport mechanisms, depending on the properties of the contaminants and the plant species involved.

Step 2: *Metabolism*: Once inside the plant, the contaminants may undergo metabolic processes facilitated by enzymes or microorganisms present in the plant's tissues. The plant may produce its own enzymes or rely on beneficial microorganisms in the rhizosphere (the soil zone influenced by the plant roots) to assist in the degradation process.

Step 3: *Transformation*: Through enzymatic reactions, the contaminants are transformed into metabolites or breakdown products. These transformation processes can involve various biochemical reactions, such as oxidation, reduction, hydrolysis, or conjugation. The transformed products are typically less toxic or more easily degraded by natural processes.

Step 4: *Storage or Release*: The transformed products may be stored within the plant tissues or released into the surrounding environment. Some plants have specialized structures or compartments, such as vacuoles, where they sequester or store the transformed contaminants. Other plants may release the transformed products into the soil or air, allowing further degradation or dispersion.

Step 4: *Rhizodegradation*: Phytodegradation can also involve the release of enzymes or compounds by plant roots into the rhizosphere, promoting the degradation or transformation of contaminants in the soil. These root exudates can enhance microbial activity and the breakdown of organic pollutants in the vicinity of the plant roots.

Uptake
(Uptake of contaminants from the soil or water)

↓

Metabolism
(Metabolic processes facilitated by enzymes or microorganisms present in the plant's tissues

↓

Transformation
(Through enzymatic reactions, the contaminants are transformed into metabolites or breakdown products)

↓

Storage or Release
(The transformed products may be stored within the plant tissues or released into the surrounding environment)

↓

Rhizodegradation
(Plant root releases enzymes or compounds by plant roots into the rhizosphere, promoting the degradation or transformation of contaminants in the soil)

Flowchart showing the mechanism of phytodegradation/rhizodegradation

In phytodegradation organic contaminants are taken up by roots and metabolized in plant tissues to less toxic substances. Phytodegradation of hydrophobic organic contaminants have been particularly successful. Poplar trees (*Populus* spp.) have been used successfully in phytodegradation of toxic and recalcitrant organic compounds. Rhizodegradation involves attenuation of organic contaminants into less toxic substances within the rhizosphere through biodegradation of soil microbes. This process is facilitated by root exudates (organic molecules) that sustain populations of soil microbes. To enhance this process a specific inocula of bacteria can be added to contaminated soils. Bacterial inocula contain strains that have desired metabolic activity to degrade targeted contaminants. Inoculating plants with genetically engineered strains of bacteria that degrade a specific contaminant has shown promising results. Also, populations of soil bacteria can be amplified using a process called biostimulation. This involves, for example, manipulating the nutrient and pH levels of the soil to increase bacterial populations.

5.4.4. Phytovolatilization

Phytovolatilization involves the uptake of contaminants by plant roots and its conversion to a gaseous state, and release into the atmosphere. This process is driven by the evapotranspiration of plants. Plants that have high evapotranspiration rate are sought after in phytovolatilization (Limmer and Burken, 2016). Transpiration is the process by which water vapors evaporate from the leaf surfaces and enter the atmosphere via stomata. Certain plant species with extensive roots systems are generally able to absorb and breakdown pollutants by producing specialized enzymes or genes. During phytovolatilization, contaminants are absorbed from the soil/water and transformed into less hazardous vapors, which are then released into the atmosphere via the plants' transpiration mechanism. The approach can be used to treat organic contaminants and some heavy metals, such as As, Se, and Hg, which exist as gaseous species in the environment. There are very few naturally occurring plants capable of turning metals into volatile forms. Phytovolatilization is a technology that uses genetically modified plants to improve their ability to volatilize metals. Furthermore, phytovolatilization causes little erosion and no disposal of contaminated plant biomass, resulting in minimal site disturbance.

The mechanism of phytovolatilization involves:

Step 1: *Uptake*: Phytovolatilization begins with the absorption of compounds from the soil by plant roots. These compounds can be organic pollutants, heavy metals, pesticides, or other substances present in the soil.

Step 2: *Translocation*: Once taken up by the roots, the compounds are transported through the plant's vascular system to the aboveground parts, such as leaves and stems. This translocation process helps distribute the compounds throughout the plant.

Step 3: *Metabolism :* Within the plant tissues, the absorbed compounds undergo various metabolic transformations. This involves enzymatic reactions that can modify the chemical structure of the compounds. The metabolism of these compounds can occur in different plant organs, including roots, stems, and leaves.

Step 4: *Conversion to VOCs:* As a result of the metabolic processes, the absorbed compounds are converted into volatile organic compounds (VOCs). VOCs are molecules that have a high vapor pressure at room temperature and can easily evaporate into the air.

Step 5: *Stomatal release*: The VOCs synthesized within the plant are released into the atmosphere through specialized structures called stomata. Stomata are small openings present on the surface of leaves, stems, and other aerial plant parts. They regulate gas exchange between the plant and its environment.

Step 6: *Diffusion*: When stomata are open, the VOCs diffuse out of the plant and into the surrounding air. The rate of diffusion is influenced by environmental factors such as temperature, humidity, and wind speed.

Uptake
(Uptake of contaminants from the soil or water)

↓

Translocation
(Metabolic processes facilitated by enzymes or microorganisms present in the plant's tissues)

↓

Metabolism
(The absorbed compounds undergo various metabolic transformations)

↓

Conversion to VOCs
(The absorbed compounds are converted into volatile organic compounds (VOCs))

↓

Stomatal release

(The VOCs synthesized within the plant are released into the atmosphere through specialized structures called stomata)

↓

Diffusion

(When stomata are open, the VOCs diffuse out of the plant and into the surrounding air)

Flowchart showing the mechanism of Phytovolatilization

Organic contaminants, especially volatile organic compounds (VOCs) are passively volatilized by plants. For example, hybrid poplar trees have been used to volatilize trichloroethylene (TCE) by converting it to chlorinated acetates and CO_2. Metals such as Se can be volatilized by plants through conversion into dimethylselenide [$Se(CH_3)_2$]. Genetic engineering has been used to allow plants to volatilize specific contaminants. For example, the ability of the tuliptree (*Liriodendron tulipifera*) to volatilize methyl-Hg from the soil into the atmosphere (as Hg0) was improved by inserting genes of modified *E. coli* that encode the enzyme mercuric ion reductase (merA).

The reaction of above mentioned examples are as below:

a. trichloroethylene (TCE) $\xrightarrow{\text{volatilization}}$ $CH_3COCl + CO_2$

b. Selenium (Se) $\xrightarrow{\text{volatilization}}$ dimethylselenide [$Se(CH_3)_2$]

c. Methyl mercury [CH_3Hg] $\xrightarrow{\text{volatilization}}$ Hg0 (monoatomic vapour)

5.5. Advantages of Phytoremediation

A significant advantage of phytoremediation is that a variety of organic and inorganic compounds are amenable to the phytoremediation process. Phytoremediation can be used either as an in situ or ex situ application. In situ applications are frequently-considered because it minimizes disturbance of the soil and surrounding environment and reduce the spread of contamination via air and waterborne wastes. It is a green technology and when properly implemented is both environmentally friendly and aesthetically pleasing to the public. Phytoremediation does not require expensive equipment or highly-specialized personnel, and it is relatively easy to implement. It is capable of permanently treating a wide range of contaminants in a wide range of environments (Farraji et al., 2016). The greatest advantage of phytoremediation is its low cost compared to conventional clean-up technologies.

5.5.1. Phytotechnology

Phytotechnology combines plant science and engineering to address technical and socio-ecological challenges. As a sector of ecotechnology, it focuses on the exploitation of specific plant-inherent processes in technical applications (Tsao 2003). These nature-based technologies have emerged as viable alternatives to conventional technologies because to their low capital costs, high success rates, low maintenance requirements, intrinsic sustainability, end use value, and aesthetic appeal.

Additional items are being prepared for the market. Various physiological plant properties are used to restore ecosystems by remediating contaminants in soil, sediments, or water (e.g., rhizofiltration, pyhtoextraction, biosorption, phytosequestration, phytohydraulics, rhizodegradation, phytovolatisation, and phytodegradation). Tapping these processes, alchemia-nova and blue carex phytotechnologies emphasize degradation properties of organic pollutants, metal accumulation, and extraction of these metals (biomining).

5.6. Disadvantages and Limitations of Phytoremediation

While phytoremediation offers several advantages, it also has certain disadvantages and limitations (Farraji et al., 2016). Here are some of them:

» ***Time-Consuming Process***: Phytoremediation is often a slow process that requires an extended timeframe to achieve significant results. It can take months to years, depending on the type and extent of contamination, plant species used, and environmental conditions. This limitation makes it unsuitable for urgent or time-sensitive remediation needs.

» ***Limited Applicability:*** Phytoremediation may not be effective for all types of contaminants. Certain pollutants, such as heavy metals with low bioavailability or persistent organic pollutants with complex chemical structures, may not be easily taken up or metabolized by plants. Additionally, some contaminants may inhibit plant growth or cause phytotoxicity, reducing the effectiveness of the process.

» ***Plant Species Selection***: The success of phytoremediation relies on selecting appropriate plant species that can tolerate the contaminants, accumulate or degrade them effectively, and thrive in the specific environmental conditions. Identifying suitable plant species for a given site and contamination type can be challenging and require extensive research and experimentation.

» ***Depth Limitations:*** Phytoremediation primarily focuses on the remediation of contaminants in the root zone of plants, limiting its effectiveness for pollutants that are present in deeper soil layers or groundwater. Contaminants that are mobile or have migrated beyond the reach of plant roots may not be effectively remediated through phytoremediation alone.

» ***Limited Scale and Site Constraints***: Phytoremediation is most effective for smaller-scale or localized contamination sites. Large-scale or widespread contamination may require extensive planting and maintenance efforts, making it logistically challenging and costly. Additionally, certain site conditions, such as steep slopes, high water tables, or extreme climates, may hinder the establishment and growth of plants, limiting the applicability of phytoremediation.

» ***Harvest and Disposal of Contaminated Biomass:*** After phytoremediation, the harvested plant biomass can contain accumulated contaminants. Proper disposal or treatment of this biomass is essential to prevent recontamination or unintended exposure to pollutants. Disposing of the contaminated biomass may require additional measures and costs.

» ***Lack of Standardization and Regulatory Framework***: Phytoremediation is still an evolving field, and there is a lack of standardized protocols and regulations for its implementation. This can lead to uncertainties in terms of performance evaluation, monitoring, and acceptance by regulatory authorities. The absence of clear guidelines and standards may hinder the widespread adoption of phytoremediation.

Despite these limitations, phytoremediation remains a valuable tool in environmental remediation, especially when combined with other techniques. Ongoing research and advancements aim to address these limitations and improve the efficacy and applicability of phytoremediation methods.

Chapter - 6

Degradation of Lignocellulosic Waste

6.1. Introduction

Lignocellulosic waste refers to biomass materials that are rich in lignocellulose, which is a complex structure found in the cell walls of plants. It includes various organic materials derived from plant sources, such as agricultural residues, forestry residues, and dedicated energy crops. Lignocellulose is the structural material consists of three main components: cellulose, hemicellulose, and lignin (Arevalo et al., 2017). Biomass in general consists of 40%–50% cellulose, 25%–30% hemicellulose, and 15%–20% lignin.

6.2. Component of Lignocellulose

6.2.1. Cellulose

Cellulose is a complex carbohydrate, specifically a polysaccharide that is found in the cell walls of plants. It is the most abundant organic compound on Earth and serves as a structural component in plant cell walls, providing rigidity and support. Cellulose fibers are long, linear chains of glucose molecules held together by hydrogen bonds, which contribute to the strength and stability of cellulose. These fibers form a network that provides structural integrity to plant tissues. Although cellulose is indigestible to humans, it plays a vital role in our diet as dietary fiber. Fiber adds bulk to our diet and helps regulate bowel movements. It also has potential health benefits, such as reducing the risk of heart disease and aiding in weight management. Cellulose is widely used in various industries due to its unique properties. It is a key component in the production of paper, textiles, and other materials. It is also used as an additive in food products, pharmaceuticals, and cosmetics. Cellulose derivatives, such as cellulose acetate and cellulose ethers, have a range of applications, including in film production, adhesives, and coatings.

DOI: 10.1201/9781003787549-6

Cellulose is a biodegradable material, which means it can be broken down by microorganisms over time. In natural environments, such as forests or composting systems, cellulose waste can decompose and contribute to nutrient cycling and soil health (Michelin et al., 2015). When cellulose waste decomposes in anaerobic conditions, such as in landfills or water bodies, it can release methane gas. Methane is a potent greenhouse gas that contributes to climate change. Proper waste management practices, such as landfill gas capture or controlled composting, can help minimize methane emissions. Improper disposal of cellulose waste can lead to its accumulation in the environment. For example, if paper and cardboard are not recycled or disposed of correctly, they can contribute to litter and pollution. This waste can clog waterways, harm wildlife, and degrade natural habitats. Cellulose waste, such as agricultural residues or wood scraps, can be utilized as a valuable resource through recycling or conversion processes. Technologies like bioenergy production, such as biomass or biofuel production, can utilize cellulose waste as a feedstock, reducing the reliance on fossil fuels. In certain cases, cellulose-based materials, such as synthetic fibers derived from cellulose, can contribute to microplastic pollution. These microplastics, formed from the degradation of synthetic textiles, can enter the environment, including water bodies, and pose potential risks to aquatic ecosystems and organisms.

6.2.2. The chemical structure of cellulose

The chemical structure of cellulose is a complex polymer composed of repeating units of glucose molecules linked together by β-1,4-glycosidic bonds. Cellulose is made up of glucose monomers, which are simple sugar molecules. Glucose has the chemical formula $C_6H_{12}O_6$. β-1,4-Glycosidic Bonds: The glucose molecules in cellulose are joined together by β-1,4-glycosidic bonds. This means that the glucose units are linked by a bond formed between the carbon-1 (C1) of one glucose molecule and the carbon-4 (C4) of the adjacent glucose molecule. The glucose units in cellulose form a long, linear chain. Each glucose molecule adds to the chain by linking to the C_4 of the previous glucose molecule. Cellulose chains are stabilized by extensive hydrogen bonding between adjacent chains. Hydrogen bonds form between the hydroxyl (OH) groups of the glucose units. This bonding creates a strong and rigid structure, contributing to the strength of cellulose. Cellulose exhibits both crystalline and amorphous regions. The crystalline regions are highly ordered and tightly packed, while the amorphous regions are less ordered and have a more random arrangement of the cellulose chains. In plants, cellulose molecules assemble into microfibrils (2–20 nm diameter and 100–40,000 nm long), which are long, parallel bundles of cellulose chains. These microfibrils provide structural support and rigidity to plant cell walls (Glasser et al., 2012).

Cellulose is also produced in a highly hydrated form by some bacteria (*Acetobacter xylinum*). Structural composition of cellulose, and arrangement of its macro and micofibrils, their arrangement of H-bonding in the cellulose molecule and its amorphous and crystalline region. Photosynthetic such as plant, algae, and some bacteria produce more than 100 million tons of organic matter each year from the fixation of carbon dioxide. Half of this biomass is made up of the biopolymer cellulose, which as a result is perhaps the most abundant molecule on the planet. The physical structure and morphology of native cellulose are complex, and fine structural details are difficult to determine experimentally. The chemical composition of cellulose is simple: the polysaccharide consists of d-glucose residues linked by β-1,4-glycosidic bonds to form linear polymeric chains of over 10,000 glucose residues.

The individual chains adhere to each other along with their lengths by hydrogen bonding and Vander Waals forces, and crystallize shortly after biosynthesis. Although highly crystalline, the structure of cellulose is not uniform. Physical and chemical evidence indicates that cellulose contains both highly crystalline and less-ordered amorphous. Although chemically simple, the extensive intermolecular bonding pattern of cellulose generates a crystalline structure that, together with other cell-wall components such as hemicellulose and lignin, results in very complex morphologies.

6.2.3. Cellulose Degradation by Aerobic and Anaerobic Microorganisms

Microorganisms including fungi, bacteria, and actinomycetes produce mainly three types of cellulase—endo-1,4-β-d-glucanase, exo-1,4-β-d-glucanase, and β-glucosidase—either separately or in the form of a complex. Most of the cellulolytic microorganisms belong to eubacteria and fungi, although some anaerobic protozoa and slime molds are able to degrade cellulose. Cellulolytic microorganisms can establish synergistic relationships with noncellulolytic species in cellulosic wastes. Most of the cellulolytic microorganisms belong to eubacteria and fungi, although some anaerobic protozoa and slime molds are able to degrade cellulose. Microorganisms capable of degrading cellulose produce a battery of enzymes with different specificities, working together. Cellulases hydrolyze the β-1,4-glycosidic linkages of cellulose.

Traditionally, they are divided into two classes referred to as endoglucanases and cellobiohydrolases. Endoglucanases (endo-1,4-β-glucanases, EGs) can hydrolyze internal bonds (preferably in cellulose amorphous regions) releasing new terminal ends. Cellulose are degraded aerobically or anaerobically by extracellular microbial enzyme cellulose, which convert cellulose into cellobiose (two glucose unit), and it can be transported (Bao et al., 2019).

6.3. Hemicellulose

The word Hemicelluloses is made up of two words Hemi and Cellulose; hemi means peripheral, and therefore hemicelluloses means "peripheral cellulose. Hemicellulose is a group of complex polysaccharides that are found in the fibers of plants along with other polysaccharides like cellulose and pectin. Hemicellulose consists of a heterogeneous group of carbohydrates where the structures of these carbohydrates are not clearly understood. Hemicelluloses are not structurally related to cellulose, nor do they contain the same building blocks but they are partially soluble in water or alkali. Hemicelluloses are mostly mixed polymers, whereas cellulose is a homopolymer of glucose (Luo et al., 2019).

Hemicelluloses are polysaccharides with β-(1 → 4)-linked sugar backbones in an equatorial orientation. It comprises xyloglucans, xylans, mannans, glucomannans, and β-(1 → 3, 1 → 4)-glucans. This form of hemicellulose is present in the cell walls of all terrestrial plants, except for β-(1 → 3, 1 → 4)-glucans, which are limited to poles and a few other taxa. Hemicelluloses include xyloglucan, xylans, mannans, and glucomannans, as well as β-(1 → 3, 1 → 4)-glucans. Some polysaccharides, such as galactans, arabinans, and arabinogalactans, are sometimes included in the hemicelluloses group. However, because these appear to be a part of pectin molecules, at least in the initial synthesis, and do not share the equatorial β-(1 → 4)-linked backbone structure, we think they should not be included in the already

All the hemicelluloses show differences in structural details between different species and in different cell types within plants. The main role of hemicelluloses is to bind cellulose microfibrils, thereby strengthening the cell wall. Hemicelluloses are synthesized by glycosyl transferases located in the Golgi membranes.

Furthermore, glycan tethering cross-links cellulose microfibrils during cellulose formation. A hemicellulose is one of numerous heteropolymers (matrix polysaccharides) found in practically all plant cell walls, along with cellulose. Although cellulose is crystalline, robust, and resistant to hydrolysis, hemicellulose is a random, shapeless arrangement with little strength. It is easily hydrolyzed by dilute acids and bases, as well as various hemicellulase enzymes. In contrast to cellulose, hemicellulose (also a polysaccharide) is made up of shorter chains—500-3000 sugar units per polymer—and has a branched structure, whereas cellulose is unbranched (Meshitsuka and Isogai, 2017). Hemicelluloses are fixed in the cell walls of plants, sometimes in chains that form a "ground"— they bind with pectin to cellulose to form a network of cross-linked fibers.

Hemicelluloses are named according to their main sugar residues in the backbone.

Xylans, consisting of d-xylose units, and glucomannans, consisting of d-glucose and d-mannose units contribute to the main hemicelluloses in hardwoods and softwoods.

Various types of Hemicellulose

Hemicellulose contains many sugar monomers. In contrast, cellulose solely includes glucose, but hemicellulose contains sugar monomers such as xylose, mannose, galactose, arabinose, and rhamnose. The majority of d-pentose sugars are found in hemicelluloses, with occasional tiny amounts of L-sugars. The sugar monomer xylose is always the most abundant, while galacturonic acid and mannuronic acid are also commonly found. Hemicellulose is a heterogeneous polymer composed of several neutral sugars, including pentoses, hexoses, and sugar acids. The most frequent hemicelluloses include xylan, xyloglucan, glucuronoxylan, arabinoxylan, and glucomannan (Brigham et al., 2018).

i. **Xylan**: Xylan is the most abundant type of hemicellulose and is primarily composed of xylose units. Xylan is the predominant hemicellulose in most plant cell walls, generally comprising about 1/3 of the total plant biomass. The composition and linkages of the side chains determine the specific variety of xylan. Removal of these side chains generally enhances the rate of degradation by endoxylanase enzymes. Xylans from grasses and annuals having an elevated level of α-l-arabinofuranoside substituents are referred to as arabinoxylans.

ii. **Mannan**: Mannan is composed of mannose units and is often found in higher amounts in softwoods. It is also present in some seeds, such as guar gum and locust bean gum. Plant mannans have β (1–4) linkages.

iii. **Glucomannan**: Glucomannan is a hemicellulosic polymer that consists of both glucose and mannose units. It is commonly found in the cell walls of some plants, including konjac root and certain types of algae. It consists of α-(1–4)-linked mannopyranose and glucopyranose backbone in a ratio of 1.6:1. Hardwood glucomannans consist of β-(1–4)-linked glucose and mannose units forming chains that are slightly branched. The ratio of mannose:glucose is about 1.5:1 or 2:1 in most hardwoods. Softwood glucomannans have occasional galactose side branches linked α-(1-6) to the mannose main chain. The α-(1-6) linkage of galactose is very sensitive to acid and alkali and maybe cleaved during alkaline extraction.

iv. **Galactomannan**: Galactomannan is composed of galactose and mannose units and is commonly found in seeds, such as guar gum and locust bean gum.

v. **Arabinoxylan**: Arabinoxylan is a hemicellulosic polymer that contains both xylose and arabinose units. It is found in various cereal grains, such as wheat, barley, and rye.

vi. **Glucuronoxylan**: Glucuronoxylan is a type of hemicellulose that contains xylose units and is modified with glucuronic acid residues. It is found in the cell walls of hardwoods and some non-woody plants. Its main constituents have α-(1–4)-xylan backbone, with 4-O-methylglucuronic acid side chains. 4-O-Methylglucuronic acid is linked to the xylan backbone by α-(1–2)-glycosidic bonds, and the acetic acid is esterified at the 2 and/or 3 carbon hydroxyl group.

Functions of Hemicelluloses in plants

Here are some key functions of hemicelluloses in plants:

» Structural Support
» Cell Wall Plasticity
» Water Absorption and Retention
» Interactions with Other Cell Wall Components
» Plant defense and Stress Response
» Seed Germination and Pollen Tube Growth

6.3.1. Microbial Degradation of Hemicellulose

Hemicelluloses are biodegraded into monomeric sugars and acetic acid. Hemicellulases are enzymes that breakdown hemicellulose and are typically classified based on how they work on various substrates. Xylan is the primary carbohydrate present in hemicellulose. Its total destruction requires the assistance of a variety of hydrolytic enzymes. It is important to distinguish between endo-α,4-β-xylanase and xylan 1,4-β-xylosidase. The former produces oligosaccharides by cleaving xylan, whereas the latter produces xylose from xylan oligosaccharides. In addition, hemicellulose biodegradation requires accessory enzymes such as xylan esterases, ferulic and p-coumaric esterases, α-l-arabinofuranosidases, and α-4-O-methyl glucuronosidases that work together to efficiently hydrolyze wood xylans and mannans. Hemicellulosic substances are degraded by diverse fungal and bacterial communities.

Bacterial Degradation of Hemicellulose

Bacteria contribute to the aerobic and anaerobic breakdown of hemicelluloses. The decomposition of hemicellulose is similar to that of cellulose in that the initial depolymerization occurs outside of the cell, and the sugars generated are then transferred inside the cell for catabolism or anabolism. Hemicellulose degradation occurs substantially more quickly than cellulose decomposition. Bacteria that degrade xylan have frequently been isolated from compost, litter, soils, sludges, the rumen system, and other gastrointestinal tracts. Some plant pathogenic Pseudomonas sp., Xanthomonas alfalfae, and Achromobacter sp. have been isolated, and their enzyme systems include hemicellulases. Hemicellulolytic Aeromonas, Bacteroides, Butyrivibrio, Ruminococcus, and Clostridium species have been found in both human and animal intestines. Plant cell walls are highly resistant to microbial destruction. The surrounding lignocellulosic components may hinder physical access to the polymer. Most xylanolytic microorganisms that grow naturally produce both cellulolytic and xylanolytic enzymes, which easily dissolve plant cell walls. Aeromonas hydrophila, Cellulomonas sp., Streptomyces olivochromogenes, Polyporous versicolor, Trichoderma harzianum, Butyrivibrio sp., Bacteroides sp., and Clostridium sp. are mannose-producing bacteria that also degrade hemicellulose. Bacterial breakdown can be aerobic or anaerobic, depending on the substrate and microbe (Mathews et al., 2015).

Aerobic Degradation

This process occurs primarily in aerobic environments, such as soil, compost, and the surface layers of natural habitats. The degradation of hemicellulose in aerobic conditions involves the action of aerobic microorganisms, including bacteria and fungi (Houfani et al., 2020). Here's an overview of the aerobic degradation process:

Enzymatic Hydrolysis: Aerobic degradation of hemicellulose begins with enzymatic hydrolysis. Enzymes produced by aerobic microorganisms, such as hemicellulases (e.g., xylanases, mannanases), break down hemicellulose into smaller sugar units, including xylose, mannose, galactose, and arabinose.

Sugar Utilization: Once the hemicellulose is hydrolyzed, the released sugar units are taken up and utilized by aerobic microorganisms as a carbon and energy source. These microorganisms can ferment the sugars, converting them into various organic compounds, such as organic acids and alcohols.

Respiration and Oxidation: The organic compounds produced from sugar fermentation are further metabolized through respiration. Aerobic microorganisms use oxygen as the final electron acceptor in their metabolic pathways, breaking down the organic compounds into simpler molecules like carbon dioxide and water. This process generates energy for the microorganisms.

CO_2 Release: During aerobic degradation, the majority of the carbon present in hemicellulose is released as carbon dioxide (CO_2) through the respiration process. CO_2 is a byproduct of the oxidation of organic compounds and is released into the atmosphere.

Mineralization: Through aerobic degradation, hemicellulose is eventually mineralized, meaning it is completely converted into inorganic compounds, such as carbon dioxide, water, and minerals. This process allows the nutrients contained in hemicellulose, such as nitrogen, phosphorus, and potassium, to become available for other organisms and for soil fertility.

Aerobic degradation of hemicellulose is essential for the breakdown of plant biomass in natural ecosystems. It contributes to the carbon cycle, nutrient cycling, and soil formation. In addition, aerobic degradation is harnessed in various industrial processes, such as composting and bioconversion technologies, to convert hemicellulose-rich biomass into useful products like biofuels, bio-based chemicals, and compost.

Anaerobic Degradation

This process occurs primarily in anaerobic environments, such as wetlands, sediments, and the digestive systems of certain animals. The degradation of hemicellulose in anaerobic conditions involves the action of specialized microorganisms, including bacteria, fungi, and archaea (Weimer, 2022). Here's an overview of the anaerobic degradation process:

Hydrolysis: The first step in anaerobic hemicellulose degradation is hydrolysis. Enzymes produced by microorganisms break down hemicellulose into smaller sugar units, such as xylose, mannose, galactose, and arabinose. These enzymes include hemicellulases, such as xylanases and mannanases.

Fermentation: After hydrolysis, the released sugar units undergo fermentation by anaerobic microorganisms. Fermentation involves the conversion of sugars into various organic compounds, such as organic acids (e.g., acetic acid, lactic acid), alcohols (e.g., ethanol), and gases (e.g., carbon dioxide, hydrogen). Different microorganisms are responsible for specific fermentation pathways.

Methanogenesis: In anaerobic environments, one of the key end products of fermentation is the production of organic acids, such as acetic acid and formic acid. Methanogenic archaea can utilize these organic acids and convert them into methane (CH_4) through a process called methanogenesis. Methane is a potent greenhouse gas and can be generated during the anaerobic degradation of hemicellulose.

Synergistic Interactions: Hemicellulose degradation in anaerobic environments often occurs in synergy with the degradation of other plant biomass components, such as cellulose and lignin. Synergistic interactions among different microorganisms enable the efficient breakdown of complex plant materials.

Volatile Fatty Acids (VFAs) Production: During anaerobic degradation, hemicellulose can be converted into volatile fatty acids (VFAs), such as acetic acid, propionic acid, and butyric acid. VFAs are important intermediates in anaerobic microbial metabolism and can serve as energy sources for other microorganisms in the ecosystem.

The anaerobic degradation of hemicellulose is significant in natural carbon cycling, nutrient recycling, and the production of biogas in anaerobic digestion processes. In anaerobic digesters, hemicellulose-rich feedstocks, such as agricultural residues and organic waste, can be converted into biogas (primarily methane and carbon dioxide) through controlled anaerobic degradation. This process is used for the production of renewable energy and the treatment of organic waste materials.

Fungal Degradation of Hemicellulose

Fungi are important in plant biomass degradation because they produce a wide range of carbohydrate-active enzymes that break down polysaccharides. Many fungal species have distinct enzyme sets for plant cell wall degradation, and understanding fungal diversity in terms of plant matter degradation is critical for improving new strains and developing enzymatic cocktails for industrial applications (Sharma and Arora, 2015). The thermophilic fungus species Thermoascus aurantiacus and Talaromyces emersonii are well-known. Endo-1,4-β-xylanase, generated by Tricoderma reesei and Aspergillus niger, cleaves endo-1,4-glycosidic bonds. Aspergillus awamori produces β-d-Xylosidase, which breaks the exo-β-1,4-glycosidic link. The white-rod fungus P. chrysosporium has been demonstrated to produce a variety of endoxylanases. The optimum temperature for fungal xylanases ranges from 40°C to 60°C. Fungal xylanases are often less thermostable than bacterial xylanases.

Actinomycetes Degradation of Hemicellulose

In thermophilic actinomycetes, three species *Thermomonospora, Actinomadura*, and mannase- producing bacteria produce enzymes that help in the degradation of hemicellulose. Xylanases from actinobacteria are active at pH 6–7. However, alkaline activity of xylanases is described from *Bacillus* sp. or *Streptomyces viridosporus*. Hemicellulose is made up of different sugars; hence, it works as a carbon source for microorganisms. Therefore, for the bioavailability of these carbons some enzymes that play a crucial role are secreted by some microorganisms.

6.4. Lignin

The term lignin comes from the Latin word lignum, which means "wood." It is found in the cell wall and provides structural stability, impermeability, and resistance to microbial attack and oxidative stress. In is one of the most prevalent organic polymers on the planet, second only to cellulose, accounting for 30% of non-fossil organic carbon and making up a quarter to a third of wood's dry mass. Lignin is an uncommon biopolymer due to its variability and lack of a definite basic structure. Its most common role is to maintain plants by reinforcing their wood (xylem cells) (Chen et al., 2018). Lignin fills the cell wall gaps between cellulose, hemicellulose, and pectin components, particularly in the tracheids, sclereids, and xylem. It is covalently bonded to hemicellulose, which crosslinks various plant polysaccharides and provides mechanical strength to the ell wall and, by extension, the plant as a whole. It is particularly plentiful in compression wood, but rare in tension wood.

Lignin plays a crucial part in conducting water in plant stems. The polysaccharide components of plant cell wall are highly hydrophilic and thus permeable to water, whereas lignin is more hydrophobic.

Lignin concentration varies according to a plant's developmental stage (Boerjan et al. 2003). For example, when plants mature, their lignin compositions become more syringyl/guaiacyl (S/G) rich. Drought, low temperature, ultraviolet irradiation, mineral shortage, mechanical injury, and disease and pest infestation all cause alterations in lignin composition. Plants, for example, have been shown to synthesize lignin that contains higher quantities of p-coumaryl alcohol near injury sites. Lignin is an amorphous heteropolymer that is insoluble in water and optically inactive. Lignin is commonly referred to as a random, complicated, irregular, heterogeneous, 3D, varyingly branching network of cross-linked phenolic (aromatic) biopolymers. Lignin is composed of three main phenylpropane units (hydroxycinnamyl alcohols that vary in their degree of methoxylation: coniferyl, sinapyl, and p-coumaryl alcohols).

Even the molecular mass of lignin is ranging from tens of thousands of Daltons to infinite. Lignins are racemic and therefore optically inactive. The racemic nature of lignins might arise from the fact that its polymerization is a nonenzymatic process.

It is made up of phenylpropane units connected together by various sorts of connections. It produces aromatic chemicals because it is a random three-dimensional network polymer made up of phenylpropanoid units with methoxyl substituents (Young and Frazer, 1987) that are linked together in random ways via ester, ether, and C-C bonds. The β-aryl-ether bond is the most prevalent inter-monomeric connection in lignin polymers. It is impossible to describe a typical structure since the lignin compound is deposited at random. It stiffens the cell wall, makes xylem vessels less permeable to water, and forms a physicochemical barrier against pathogenic microbial attack. The polysaccharides component of plant cell wall is highly hydrophilic and thus permeable to water, whereas lignin is more hydrophobic. The cross linking of polysaccharides by lignin is an obstacle for water absorption into the cell wall. It is a heterogeneous phenolic polymer and plays crucial roles in the development and physiology of vascular plant.

Lignin is found in cell wall in a complex with cellulosic and hemicellulosic polysaccharides providing protection to these carbohydrates from biological degradation. Additionally, the lignin in paper is an inhibiting agent in paper degradation (in landfills).

Lignin also provides mechanical rigidity enabling plants to defy gravitational forces and grow skyward, affording their sometimes immense size and volume (such as giant redwoods, sequoias). Lignin enables vascular integrity and protects the cell wall polysaccharides, which makes it difficult to exploit the sugars for biofuels. Lignin offer protection against pathogens, pests, and natural or mechanical wounding.

The vital properties of lignin are evidenced from the impaired growth and viability of plants with low or disorganized lignin (naturally or due to genetic engineering). Lignin enables the transport of water from the roots to the leaves through the xylem vasculature. The lignified cell walls are essentially dead cells linked together to form long hollow tubes. The inside of these tubes are the sites for lignin deposition (i.e., lignification). Lignin is deposited in the secondary walls and among the fibers amid nonlignified tissues. Such polydisperse deposition also hinders the isolation and characterization of pure, intact lignin. It also plays a significant role in carbon cycle, formation of humus, and increases the photosynthetic productivity of plant by providing cation exchange capacity in the soil and expending the capacity of moisture retention between flood and drought condition by its hydrophilic nature.

6.4.1. Synthesis of lignin and its precursor

Lignin formation is a complex process; it begins in the cytosol with the synthesis of glycosylated monolignols from the amino acid phenylalanine. Lignin is formed by radical polymerization of guaiacyl (G) units from precursor coniferyl alcohol, syringyl (S) units from precursor sinapyl alcohol, and p-hydroxyphenyl (H) units from precursor p-coumaryl alcohol. Polymerization occurs in the cell walls after polysaccharide deposition and is initiated by peroxidase- or laccase-mediated one electron oxidation of phenylpropanoid precursor to phenoxy radicals, which undergoes coupling to form a complex, cross-linked lignin polymer. The final result of this polymerization is a heterogeneous structure whose basic units are linked by C–C and aryl–ether linkages, with aryl-glycerol β-arylether being the predominant structure (Liu et al., 2019).

p-Coumaric acid undergoes the phenylpropanoid pathway in order to generate the three monolignols with varying methoxy substitutions. It begins with the enzymatic deamination of phenylalanine by phenylalanine ammonia lyase. After monolignol synthesis they are transported to lignification sites during wall development. Glucosyl transferases and β-glucosidases regulate the availability of monolignols during lignification. Cell wall-associated β-glucosidases might release the aglycone (monolignols) for subsequent polymerization during lignin synthesis.

Once the monolignols are transported to lignification sites, they undergo radicalization. Redox enzymes such as peroxidases, phenol oxidases, and laccases have been implicated in the oxidative radicalization of the monolignols. Redox shuttle mediators involve monolignols in the radicalization of monolignols for the polymerization of lignin. Laccases utilize oxygen for oxidation, whereas peroxidases utilize hydrogen peroxide (H_2O_2) for oxidizing the monolignols. Cellwall-associated peroxidases and laccases might additionally enable lignin polymerization at specific sites.

6.4.2. Microbial Degradation of Lignin

The degradation of lignin to enhance the release of fermentable sugars from plant cell walls presents a challenge for biofuel production from lignocellulosic biomass. The discovery of novel lignin-degrading enzymes from bacteria could provide advantages over fungal enzymes in terms of their production and relative ease of protein engineering (Kamimura et al., 2019). Lignin degradation is a slow process, which takes many years for degradation naturally. To remove lignin from the environment, many studies had been conducted on microorganisms by many researchers.

Bacterial Biodegradation

Bacterial degradation of lignin is a complex process in which certain bacteria have the ability to break down lignin, a complex and highly resistant polymer found in the cell walls of plants. Lignin degradation is an essential step in the carbon cycle and plays a crucial role in the recycling of plant biomass. There are several bacterial species known to possess lignin-degrading capabilities. One well-studied example is the bacterium *Rhodococcus jostii* RHA1, which has been extensively investigated for its lignin degradation abilities. Other bacteria such as *Pseudomonas, Sphingomonas*, and *Streptomyces* species have also been found to have lignin-degrading enzymes. The degradation of lignin by bacteria involves the action of various enzymes called ligninases or ligninolytic enzymes. These enzymes can be categorized into three main types: lignin peroxidases (LiPs), manganese peroxidases (MnPs), and laccases. Each of these enzymes has a specific role in breaking down different components of lignin. Lignin peroxidases are the most well-known lignin-degrading enzymes and are produced by several bacterial species. They are capable of oxidizing lignin by breaking down its aromatic rings, resulting in the formation of smaller molecules. Manganese peroxidases, as the name suggests, require manganese ions for their activity and can also break down lignin. Laccases, on the other hand, are copper-containing enzymes that can oxidize lignin and other aromatic compounds. Bacterial degradation of lignin is often a collaborative process involving multiple species working together. Some bacteria may specialize in breaking down certain components of lignin, while others can utilize the degradation products produced by their counterparts. The bacterial degradation pathway of lignin involves a series of enzymatic reactions that break down the complex lignin polymer into smaller, more manageable fragments.

While the exact pathway can vary among different bacteria, the overall process typically involves the following steps:

Activation: Bacteria that possess lignin-degrading capabilities produce specific enzymes, such as lignin peroxidases, manganese peroxidases, and laccases, which are involved in lignin degradation. These enzymes are typically induced when the bacteria are exposed to lignin or lignin-related compounds.

Initial attack: The lignin polymer is initially attacked by ligninolytic enzymes. Lignin peroxidases and manganese peroxidases act on the lignin structure, breaking down the complex aromatic rings and generating reactive intermediates.

Cleavage of lignin side chains: Bacteria produce enzymes, such as DyP-type peroxidases and auxiliary activities (AA) family enzymes, which target the side chains of lignin molecules. These enzymes can cleave ester bonds, ether linkages, and other chemical bonds present in the lignin structure, resulting in the release of various lignin-derived compounds.

Aromatic ring degradation: Bacteria possess enzymes, such as extradiol and intradiol ring-cleaving dioxygenases, that can further break down the aromatic rings released from lignin. These enzymes catalyze the cleavage of the aromatic rings, generating smaller aromatic compounds that can be metabolized by the bacteria.

Utilization of lignin-derived compounds: Bacteria have metabolic pathways to utilize the smaller lignin-derived compounds produced during lignin degradation. These compounds can be further metabolized through various biochemical reactions, ultimately providing a source of carbon and energy for the bacteria.

It is important to note that lignin is a complex and heterogeneous polymer, and the degradation pathway can vary depending on the specific bacterial species and the environmental conditions. Additionally, bacterial consortia or communities often work together to efficiently degrade lignin, with different bacteria specializing in different steps of the degradation process.

Fungal Biodegradation

Fungi are the only microorganisms studied extensively for the degradation of lignin. Lignin degradation by fungi has been discussed by a number of researchers. All fungi produced LiP, MnP, and laccase except *Daedalea flavida* and *Dichomitus squalens*, which lack LiP and MnP, respectively. Lignin peroxidases are particularly important in the initial stages of lignin degradation. They can directly oxidize lignin, generating highly reactive free radicals that attack the lignin polymer. Manganese peroxidases and laccases are also involved in lignin breakdown, but their specific roles are still not fully understood (Madadi and Abbas, 2017).

Once lignin is degraded, the remaining products can be further broken down by other fungal enzymes, such as cellulases and hemicellulases, which degrade the polysaccharides present in the plant cell walls. The ability of fungi to degrade lignin has significant implications in various industries. For example, it can be harnessed in the field of biotechnology for the production of biofuels and other value-added products from lignocellulosic biomass. Understanding the mechanisms of lignin degradation by fungi can help in the development of more efficient and cost-effective processes for biomass utilization. It is worth noting that not all fungi have the ability to degrade lignin. Some fungi, known as brown-rot fungi, primarily degrade the cellulose component of wood, leaving behind a brown, degraded residue. Others, such as soft-rot fungi, focus on breaking down the hemicellulose component of wood. The specific fungal species and environmental conditions play a significant role in determining the extent and efficiency of lignin degradation.

6.5. Mechanism of Ligninolytic Enzymes for Degradation of Lignocellulosic Waste

Ligninolytic enzymes play a vital role in lignin degradation, bioremediation, paper and pulp industry, textile and dye industry, biomass conversion, and biofuel production. Their versatility and ability to break down complex organic compounds make them valuable tools in various industrial and environmental applications. The most common lignocellulosic waste degrading enzymes are Laccase (EC 1.10.3.2), Lignin peroxidase (EC 1.11.1.14), Manganese peroxidase (EC 1.11.1.13) and versatile peroxidase (EC 1.11.1.16).

6.5.1. Laccase: structure and properties

Laccase is a copper-containing oxidase enzyme that is found in various organisms, including fungi, plants, and bacteria (Alcalde, 2007). It plays a crucial role in catalyzing the oxidation of a wide range of substrates, including phenolic compounds and aromatic amines. The structures of laccase consist of:

Copper-binding sites: Laccase contains multiple copper-binding sites that are essential for its catalytic activity. These sites are typically classified as Type 1, Type 2, and Type 3 copper centers. The Type 1 copper site is a mononuclear copper center coordinated by cysteine and histidine residues. The Type 2 and Type 3 copper sites are binuclear copper centers involved in the catalytic redox reactions.

Multicopper oxidase (MCO) domain: Laccase has a conserved multicopper oxidase (MCO) domain, which is responsible for the binding and coordination of copper ions. This domain contains several conserved amino acid residues involved in copper coordination and catalysis.

Four-copper center architecture: Laccase typically exhibits a four-copper center architecture, consisting of one Type 1 copper center and a trinuclear copper cluster composed of one Type 2 and two Type 3 copper centers. This arrangement allows laccase to undergo a four-electron reduction and oxidation cycle during its catalytic reactions.

Glycosylation: Laccase proteins are often glycosylated, meaning they have sugar molecules attached to specific amino acid residues. Glycosylation can affect the stability, activity, and solubility of laccase and it may also play a role in protein folding and protection against proteolytic degradation.

Structural variability: Laccase structures can vary among different organisms and even among different isoforms within the same organism. While the overall fold and copper-binding sites are conserved, there can be variations in the arrangement of secondary structure elements, loop regions, and surface features.

Determining the precise three-dimensional structure of laccase has been challenging due to its large size, multiple copper centers, and structural flexibility. However, advancements in X-ray crystallography, nuclear magnetic resonance (NMR) spectroscopy, and cryo-electron microscopy (cryo-EM) have provided valuable insights into the structure-function relationships of laccase enzymes from various sources. These studies have contributed to a better understanding of the catalytic mechanisms and have paved the way for engineering laccases with improved properties for biotechnological applications.

6.5.1.1. Properties of laccase enzyme

i. *Oxidative activity*: Laccase exhibits strong oxidative activity, catalyzing the oxidation of a wide range of substrates. It can oxidize various phenolic compounds, aromatic amines, and other organic substrates. This property allows laccase to participate in diverse biocatalytic reactions, including the degradation of lignin, detoxification of pollutants, and transformation of industrial chemicals.

ii. *Broad substrate specificity*: Laccase shows broad substrate specificity, meaning it can act on a wide variety of compounds. It can oxidize both natural and synthetic substrates, including lignin, dyes, pigments, phenols, aromatic compounds, and certain drugs. This versatility makes laccase useful in various industrial processes, such as bioremediation, textile dyeing, and production of high-value chemicals.

iii. *pH and temperature stability*: Laccase exhibits stability over a wide range of pH and temperature conditions. Different laccase isoforms may have varying pH optima, but they generally function effectively in the neutral to slightly acidic pH range. Laccase enzymes derived from extremophilic organisms can also withstand high temperatures. This stability allows laccase to be used in different biotechnological applications under diverse environmental conditions.

iv. *Copper dependence*: Laccase is a copper-dependent enzyme, meaning it requires copper ions for its activity. The copper ions are essential for the formation and stabilization of the copper-binding sites within the enzyme's structure. These copper centers play a crucial role in the redox reactions catalyzed by laccase, enabling electron transfer between the enzyme and its substrates.

v. *Resistance to inhibitors:* Laccase has shown relative resistance to certain inhibitors, including heavy metals, organic solvents, and various chemical compounds. This resistance is attributed to the structural stability of laccase and its ability to maintain catalytic activity in the presence of inhibitory substances. It allows laccase to be used in environments or processes where the presence of inhibitory compounds might otherwise hinder enzyme activity.

vi. *Environmental compatibility*: Laccase is considered an environmentally friendly enzyme due to its natural origin and its potential to replace or reduce the use of harsh chemical oxidants in various industrial processes. It offers a greener alternative for applications such as bioremediation, pulp bleaching, dye decolorization, and wastewater treatment.

vii. *Enzyme engineering potential*: Laccase has been subjected to extensive research in enzyme engineering and protein design. Through methods like directed evolution and protein engineering techniques, it is possible to modify and optimize laccase properties for specific applications, such as enhanced catalytic activity, substrate specificity, thermal stability, or pH tolerance.

The properties of laccase make it a versatile biocatalyst with significant potential in various industries, including biotechnology, environmental remediation, pulp and paper, textile, pharmaceuticals, and bioenergy. Ongoing research and advancements in enzyme engineering continue to expand the scope of laccase applications and improve its overall performance.

6.5.2. Lignin peroxidase: structure and properties

Lignin peroxidase (LiP) is an enzyme that belongs to the class of heme peroxidases and is primarily produced by white-rot fungi. It plays a key role in lignin degradation by catalyzing the oxidative depolymerization of lignin. The structure of lignin peroxidase is characterized by several important features (Piontek et al., 2001):

Heme-containing active site: LiP contains a heme prosthetic group that serves as the catalytic center of the enzyme. The heme group consists of an iron ion coordinated by a porphyrin ring. It participates in redox reactions by accepting and donating electrons during the catalytic process.

Secondary structure elements: LiP typically consists of a predominantly alpha-helical structure with some beta-sheet regions. The secondary structure elements contribute to the overall stability and folding of the enzyme.

Active site residues: The active site of LiP includes specific amino acid residues that interact with the heme group and participate in the catalytic reaction. These residues are involved in substrate binding, electron transfer, and stabilization of reaction intermediates.

Calcium-binding sites: LiP often contains calcium-binding sites that play a role in stabilizing the protein structure and maintaining its overall integrity. Calcium ions help in enzyme folding and are known to enhance the activity and stability of LiP.

Disulfide bonds: Disulfide bonds can be present within the LiP structure, contributing to the stability and correct folding of the enzyme. These bonds are formed between cysteine residues and help maintain the structural integrity of the protein.

Variable regions: The overall structure of LiP can exhibit some variability, particularly in the loops and surface regions. These variable regions can contribute to substrate specificity and interaction with other proteins or molecules.

6.5.2.1. Properties of Lignin peroxidase

i. *Lignin degradation*: LiP plays a crucial role in lignin degradation by catalyzing the oxidative depolymerization of lignin. It can break down complex aromatic structures and generate reactive intermediates that can be further metabolized by other enzymes. This property enables the efficient breakdown of lignin, a complex and recalcitrant polymer, into smaller molecules that can be utilized by organisms or used for the production of value-added chemicals and biofuels.

ii. *Versatile substrate specificity*: LiP exhibits broad substrate specificity and can act on a variety of substrates, including lignin-derived compounds, synthetic dyes, aromatic pollutants, and other phenolic compounds. This versatility allows LiP to participate in diverse biocatalytic reactions, including the detoxification of environmental pollutants, decolorization of industrial wastewater, and modification of lignocellulosic biomass for various applications.

iii. *Oxidative activity*: LiP is an oxidative enzyme that utilizes hydrogen peroxide (H2O2) as a co-substrate to drive its catalytic reactions. It can catalyze the oxidation of both low-molecular-weight phenolic compounds and high-molecular-weight lignin polymers. This oxidative activity enables LiP to initiate the breakdown of lignin and other recalcitrant compounds by generating free radicals and reactive intermediates.

iv. *Calcium dependence*: LiP activity is often dependent on the presence of calcium ions. Calcium ions are required for the correct folding, stability, and activity of the enzyme. They play a role in maintaining the tertiary structure and catalytic properties of LiP, and their absence can significantly affect enzyme function.

v. *Redox cycling*: LiP participates in redox cycling reactions by accepting electrons from lignin or other substrates and transferring them to the heme cofactor. This cycling allows LiP to sustain its oxidative activity and perform multiple catalytic cycles.

vi. *pH and temperature stability*: LiP exhibits stability over a wide range of pH and temperature conditions. Different LiP isoforms may have varying pH optima, but they generally function effectively in the acidic to neutral pH range. LiP enzymes derived from thermophilic organisms can also withstand higher temperatures. This stability enables LiP to be used in various industrial processes under diverse environmental conditions.

vii. *Potential for protein engineering*: LiP has been the subject of protein engineering studies aimed at enhancing its catalytic activity, stability, and substrate specificity. Through techniques such as directed evolution and site-directed mutagenesis, it is possible to modify and optimize LiP properties for specific applications or to engineer variants with improved performance.

6.5.3. Manganese peroxidase: structure and properties

Manganese peroxidase (MnP) is an enzyme produced by white-rot fungi and some bacteria. It plays a significant role in lignin degradation by catalyzing the oxidation of manganese ions (Mn^{2+}) and other substrates (Gold et al., 2000). The structure of MnP is characterized by several important features:

Heme-containing active site: MnP contains a heme prosthetic group that serves as the catalytic center of the enzyme. The heme group consists of an iron ion coordinated by a porphyrin ring. Similar to other peroxidases, the heme group in MnP participates in redox reactions by accepting and donating electrons during the catalytic process.

Calcium-binding sites: MnP often contains calcium-binding sites that contribute to the stability and proper folding of the enzyme. Calcium ions play a role in enzyme structure, stability, and activity. They are known to enhance the thermal stability and activity of MnP.

Manganese-binding sites: MnP has specific sites for binding manganese ions (Mn2+). These binding sites facilitate the oxidation of Mn2+ to Mn3+, which is a crucial step in the catalytic cycle of MnP. The oxidation of Mn2+ generates reactive intermediates that are involved in lignin degradation.

Secondary structure elements: MnP typically exhibits a mixed alpha-helical and beta-sheet structure. The secondary structure elements contribute to the overall stability and folding of the enzyme.

Variable regions: Like other peroxidases, MnP can have variable regions, such as loops and surface regions, which may contribute to substrate specificity and interaction with other molecules or proteins.

6.5.3.1. Properties of Manganese peroxidase

i. *Lignin degradation*: MnP plays a crucial role in lignin degradation by catalyzing the oxidation of lignin and related compounds. It is particularly effective in breaking down lignin structures with high redox potential, such as phenolic compounds. MnP generates reactive intermediates that can initiate lignin depolymerization, leading to the breakdown of complex lignin molecules into smaller fragments.

ii. *Broad substrate specificity*: MnP exhibits broad substrate specificity, allowing it to act on a wide range of compounds, including lignin-derived substrates, dyes, aromatic pollutants, and phenolic compounds. This versatility makes MnP useful in various biocatalytic reactions, including bioremediation, wastewater treatment, and modification of lignocellulosic biomass for industrial applications.

iii. *Oxidative activity*: MnP is an oxidative enzyme that utilizes hydrogen peroxide (H2O2) as a co-substrate to drive its catalytic reactions. It catalyzes the oxidation of Mn2+ ions, which leads to the generation of reactive oxygen species (ROS) and other free radicals. These reactive species are involved in the degradation of lignin and other substrates through oxidation reactions.

iv. *Synergistic action with other enzymes*: MnP often works in synergy with other ligninolytic enzymes, such as lignin peroxidase (LiP) and laccase, to enhance lignin degradation efficiency. The combined action of these enzymes enables the complete breakdown of lignin, as they have complementary substrate specificities and catalytic mechanisms.

v. *pH and temperature stability*: MnP exhibits stability over a broad range of pH and temperature conditions. Different MnP isoforms may have varying pH optima, but they generally function effectively in the slightly acidic to neutral pH range. MnP enzymes derived from thermophilic organisms can also withstand higher temperatures. This stability allows MnP to operate under diverse environmental conditions.

vi. *Potential for protein engineering*: MnP has been the subject of protein engineering studies aimed at improving its catalytic efficiency, substrate specificity, and stability. Techniques such as directed evolution and site-directed mutagenesis can be employed to modify and optimize MnP properties for specific applications or to develop variants with enhanced performance.

vii. *Environmental compatibility*: MnP is considered environmentally friendly due to its natural origin and its potential to replace or reduce the use of harsh chemical oxidants in various industrial processes. It offers a greener alternative for applications such as bioremediation, pulp bleaching, textile dyeing, and wastewater treatment.

6.6. Bacterial degradation studies of Lignocellulosic waste in laboratory for environment safety

Lignocellulosic waste, which is known rich in lignin, cellulose, and hemicellulose, can pose several environmental problems if not managed properly. Pulp and paper mill industry effluent is major source of Lignocellulosic wastes in the environment. Most of the studies have been conducted for the biodegradation of Lignocellulosic waste by fungi but the biodegradation of Lignocellulosic waste by potential bacteria has been reported least on laboratory scale. Hence, the case study of the biodegradation of organic pollutant detected from the pulp paper mill effluent by the potential ligninolytic bacterial strain is given below.

Step 1: Isolation and purification of bacterial strains

Step: 2 Screening of isolated bacterial strain for ligninolytic activity

By using a plate assay method, the pure bacterial strains were screened for diverse nature of ligninolytic enzyme activity, such as manganese peroxidase (MnP), lignin peroxidase (Lip), and laccase (Lac) activity. For the screening, MSM with various substrates was utilised. Phenol red (0.1%) was used as a substrate for manganese peroxidase, while Azure B was used as a substrate for lignin peroxidase (0.002%). In contrast, laccase activity was identified in Nutrient Agar (NA) medium that contained $CuSO_4$ (1 mM), beef extract (3.5), peptone (5.0), NaCl (5.0), and guaiacol as substrate (in gm/l). The change of dark reddish pink to yellow colour indicated the presence of MnP activity, and the disappearance of blue colour of Media (i.e. Azure B) indicated the production of LiP from bacteria. Similarly, the presence of brown colour halos indicated the positive test laccase activity.

Step: 3 Assessment of lignocellulosic compounds degradation by isolated bacterial strains

To detect the metabolites present in pulp paper mill effluent the sample was processed for HPLC analysis. For confirmatory analysis, 100 mL of bacterial degraded and undegraded sample were centrifuged. Supernatants were acidified with the help of HCl with consistent agitation and extracted thrice with 3volume of extraction solution (ethyl acetate/methanol/acetonitrile) acetate. The organic layer was pooled and dewatered over anhydrous $Na_2SO_{4.}$ Residues were dried under a stream of nitrogen gas and dissolved in acetonitrile for HPLC analysis. The chromatogram and detected organic pollutant before and after bacterial degradation is shown in Table 6.1.

Step 4: Detection of lignocellulosic compound and their degradation of metabolic products

Further, to detect the lignocellulosic compound and their degradation of metabolic products, Gas chromatography–mass spectrometry (GC-MS) was performed. The method is as followed: Ethyl acetate solvent was used to recover organic pollutants from the distillery and rhizospheric sludge, and it was used to detect a wide spectrum of organic compounds at a slightly alkaline pH (i.e. 8.10).

An amount of 25.0 ml pulp paper mill effluent sample of control and after bacterial treatment was weighed and placed in a 250 mL Erlenmeyer flask. Subsequently, 25 mL ethyl acetate was added to each flask individually and vigorously mixed to create a homogenised solution for the solubility of organic components. The samples were also subjected to vortex agitation (for 3 min), sonication (3 min on, 30 s off, three times), and centrifugation for 10 min at 5000 rpm. To ensure complete extraction from sample, the extraction procedure was repeated thrice.

After the procedure was finished, the organic layer that had been separated was collected from the separating funnel and dehydrated by passing through a Na_2SO_4 column. The dried organic extracted precipitate was passed through a 0.22 m syringe filter after being diluted in 1.5 mL of ethyl acetate solution.

Table 6.1. Detection of comparative list of various organic compounds by GC-MS extracted from pulp paper mill effluent control and treated sample

RT	Compounds	Control	Degraded
10.08	Benzoic acid	+	-
10.11	Beta hydroxybutyric acid	+	+
10.91	3,4-Dimethylbenzaldehyde	+	+
12.55	Tetradecanoic acid	+	-
12.99	2-Methoxy-4-vinylphenol	+	-
13.45	Benzene propionic acid	-	+
13.80	3-Chloropropionic acid	+	-
14.70	3-Hydroxy-4- methoxymandelic acid	-	+
15.74	5,6-Dimethoxy-1-indanone	+	-
15.76	4-Ethoxy benzoic acid	+	-
16.51	Dibutyl phthalate	-	+
16.67	N-hexadecanoic acid	-	+
17.17	Pyrene	+	-
19.21	Octadecanoic acid	+	-
20.11	4- Hydroxy phenol	+	-
21.57	Bis (2-ethylhexyl) phthalate	+	-
21.72	Phthalic anhydride	-	+
23.30	Benzyl butyl phthalate	+	-
24.86	4,4-Diethyl-2,5-octadiyne	+	-
26.06	Propionic Acid	-	+

Chapter - 7

Heavy Metal Toxicity in the Environment and Its Bioremediation

7.1. Introduction

Heavy metals are considered as major source of environmental contamination due to their toxic nature and their ability to accumulate. Industrialization, urbanization, and agricultural activities resulted in increase of heavy metal concentrations in different habitats compared to their natural background levels. Street dust, roadside soil and plants growing in these polluted areas are subjected to receive high amount of heavy metals both from hazardous gases emitting from motor vehicle and transported toxic materials (Tchounwou et al., 2012). Vehicular activities on roads by motors also promote metal levels especially lead and nickel, in our ecosystem through burning fossil fuels and vehicle wear i.e., brakes, vehicle body, tyres and vehicular fluids. Lead is the most prevalent heavy metal contaminant found in the aquatic environments and nearby soils of industrial region.

Heavy metals are dangerous because they tend to bioaccumulate. Bioaccumulation means an increase in the concentration of a chemical in a biological organism over time, compared to the chemical's concentration in the environment. Compounds accumulate in living things any time they are taken up and stored faster than they are broken down (metabolized) or excreted. The toxicity of metals depends on factors such as their chemical form, concentration, and exposure duration. Environmental regulations and proper waste management practices are essential for minimizing the release and impact of metal xenobiotics on ecosystems and human health.

7.2. Metal Toxicity to Plants

Heavy metals like As, Hg, Pb, or Cd play no role in the growth of plants as they have no known physiological function in plants. Other heavy metals like Co, Cu,

DOI: 10.1201/9781003787549-7

Zn, Mo, and Fe are essential for the growth of the plant; if the concentration rises above the optimal concentration, it may lead to poisoning. Plants are an integral part of the food chain, and considering the edible parts, there is a higher risk of transferring heavy metals to various other levels of the food chain; hence, it is a matter of concern. Plants absorb heavy metals mainly through their roots. Absorption and accumulation of metals are dependent on several factors, such as temperature, pH, and moisture. Different seasons or, to be precise, different temperatures drive the absorption and accumulation of heavy metals; for example, in the summer season, Cd, Zn, and Cr uptake and accumulation in spinach were higher, and in the winter, levels of Cu, Ni, and Pb were found to be higher (Angulo et al., 2021).

Plant species determine the extent of toxic metal accumulation. An increase in levels of Pb in the soil can reduce its productivity, whereas a reduced amount of Pb may hinder several important plant processes, like photosynthesis and mitosis. Symptoms of toxicity can be like wilting of leaves and short brown roots. Heavy metals make plants weak, blocking their growth, and metabolic disorders are also seen. Heavy metals also reduce the ability of leguminous plants to fix molecular nitrogen. An increased level of Pb also delays the process of germination.

7.3. Effects of Metals on the Aquatic Environment

Heavy metals, even at very low concentrations, might be highly persistent and toxic and have the potential to induce serious stress in aquatic organisms. Heavy metals have high signifcance in the context of ecotoxicology. Most importantly, if metal contaminants are not degraded by microbes, once they enter a marine ecosystem, they stay there forever. Heavy metal-contaminated rivers can cause devastating effects as they not only damage the ecological balance but also affect the diversity of organisms in that river (Aziz et al., 2023). A contaminated river poses a great risk not only to aquatic organisms but also to other organisms that are dependent on that river as their primary water source.

Heavy metals that are introduced into aquatic systems are usually attached to particulate matter that eventually gets settled and incorporated into sediments. Aquatic organisms and rooted aquatic macrophytes take up the pollutants that are sediment-bound. The majority of trace elements that get introduced into the aquatic environment are bound with bottom sediments, and organisms in or around the sediment get affected adversely. In the aquatic system, contaminants have heavy metals in them that have the potential to stimulate the production of ROS, which can harm fish and other organisms. Out of all the contaminants, mercury is an important one as it harms marine life and has the potential to harm humans too. In aquatic sediments, methyl mercury is formed by the bacterial methylation of organic mercury, and interestingly, traces of mercury found in fish muscles are mostly methyl mercury. Macro invertebrates that are affected by toxic heavy metals are yet to be assessed. The effect of heavy metals on drift-prone macro invertebrates is important to study because they have an important commercial value, and several salmonid species also depend on them.

7.4. Effects of Metals on Human

Some metals are vital for the existence of humans, such as iron, magnesium, copper, zinc, and, at specific concentrations, but these can also be highly lethal to humans as well as animals at elevated concentrations. Millions of individuals are regularly exposed to various toxic metals via air, soil, and water (Fu and Xi, 2020). External respiration, contaminated water, and food with toxic metals might be the major ways to enter biological systems.

Daily intake of low concentrations of metal ions shows acute and chronic health effects in animals. The toxicity levels of metal ions are dependent on the exposure, occurrence, duration, biological system, age, gender, individual, and dietary features. Some specific effects of different heavy metals on human health:

i. Lead (Pb):

» Neurological effects: Lead exposure can cause intellectual and behavioral disabilities in children. It can lead to learning difficulties, lowered IQ, developmental delays, and attention deficit hyperactivity disorder (ADHD).

» Hematological effects: Lead can interfere with the production of red blood cells, leading to anemia.

» Renal effects: Prolonged lead exposure can cause kidney damage and impaired renal function.

» Cardiovascular effects: Lead exposure has been associated with increased blood pressure, heart disease, and an elevated risk of cardiovascular events.

ii. Mercury (Hg):

» Neurological effects: Mercury is highly toxic to the nervous system. It can cause tremors, memory loss, cognitive impairment, irritability, and difficulties with coordination and motor function. Severe mercury poisoning can lead to hallucinations and even coma.

- » Developmental effects: Prenatal exposure to mercury can harm the developing nervous system, leading to cognitive and developmental impairments in children.
- » Renal effects: Mercury can damage the kidneys and impair their function.

iii. Cadmium (Cd):

- » Renal effects: Cadmium is primarily toxic to the kidneys. Prolonged exposure can cause kidney damage, renal dysfunction, and an increased risk of kidney disease.
- » Skeletal effects: Cadmium can accumulate in bones and interfere with calcium metabolism, leading to osteoporosis and increased fracture risk.
- » Respiratory effects: Inhalation of cadmium fumes or dust can cause lung damage, chronic bronchitis, and potentially lung cancer.
- » Reproductive effects: Cadmium exposure has been associated with adverse effects on male and female reproductive systems, including infertility and decreased fertility.

iv. Arsenic (As):

- » Dermatological effects: Arsenic exposure can cause skin lesions, discoloration, and increased risk of skin cancer.
- » Carcinogenic effects: Arsenic is a known human carcinogen, associated with an increased risk of various cancers, including skin, lung, bladder, and kidney cancer.
- » Cardiovascular effects: Chronic arsenic exposure has been linked to cardiovascular diseases such as atherosclerosis, hypertension, and heart attacks.
- » Respiratory effects: Inhalation of arsenic-containing dust or fumes can irritate the respiratory system and lead to respiratory problems.

v. Chromium (Cr):

- » Dermatological effects: Hexavalent chromium (Cr(VI)) compounds are strong skin sensitizers and can cause allergic dermatitis.
- » Respiratory effects: Inhalation of Cr(VI) compounds can lead to respiratory issues, including asthma, bronchitis, and lung cancer.

» Carcinogenic effects: Hexavalent chromium is a known human carcinogen and has been associated with an increased risk of lung cancer.

7.5. Bacterial Bioremediation of Metals

A wide range of heavy metal ions can be reduced by bacteria. As a result of this adaptation, biological components developed defenses against heavy metal ions. Bacterial tolerance to metals may be infuenced by a number of circumstances, such as the transfer of metal ions into the cell, efux, both extracellular and intracellular sequestration, and redox reactions (Sreedevi et al., 2022). The extracellular barrier may serve as protection against the entry of metal ions. But using ionizable groups in the cell wall or capsule, bacteria may absorb metal ions. In order to export metal ions from cells, bacteria use active transport and efux mechanisms. Then, metals combine with various compounds (metallothioneins, phytochelatins) in the cell cytoplasm or accumulate via cellular components (periplasmic proteins, outer membrane proteins) in the periplasm or the outer membrane. Certain bacteria can produce energy by using metals and metalloids as electron acceptors or donors. Metals in their oxidized state may be used as terminal electron acceptors in bacterial anaerobic respiration. Enzymatic reduction would also result in the formation of less toxic heavy metal forms.

Bacteria can bioremediate heavy metals in the environment through various mechanisms. Here are some key ways in which bacteria contribute to the remediation of heavy metal contamination:

7.5.1. Bioaccumulation: Certain bacteria have the ability to accumulate heavy metals from their surroundings. They possess metal transport proteins or channels on their cell membranes that facilitate the uptake of metals. Once inside the bacterial cells, metals can accumulate in high concentrations due to active transport mechanisms or passive diffusion. The detail process involves:

(i) *Extracellular binding*: Bacteria can produce extracellular polymeric substances (EPS) that contain functional groups, such as carboxyl, amino, or phosphate groups, which have a high affinity for metal ions. These functional groups can bind and sequester metal ions from the surrounding environment, effectively immobilizing them and preventing their mobility.

(ii) *Intracellular uptake*: Bacteria possess transport systems, such as specific metal ion uptake channels or transporters, which enable the uptake of metal ions from the extracellular environment into the bacterial cell. These transport systems are often highly specific to particular metal ions, allowing bacteria to selectively accumulate specific metals.

i. *Metal reduction and precipitation*: Some bacteria have the ability to reduce metal ions to a less toxic and less soluble form, leading to their precipitation or immobilization. For example, certain bacteria can reduce soluble toxic forms of metals, such as hexavalent chromium (Cr(VI)) or mercury (Hg (II)), into less soluble and less toxic forms, such as Cr(III) or elemental mercury (Hg^0). The precipitated metals can then be either stored intracellularly or deposited extracellularly.

ii. *Biovolatilization:* Certain bacteria have the capability to transform metal ions into volatile forms, which can be released into the atmosphere. This process, known as biovolatilization, can be employed to remove volatile metals from contaminated sites. For example, some bacteria can convert mercury (Hg) into volatile methylmercury (MeHg), which can then be released into the air.

iii. *Biosorption*: Biosorption involves the binding of metal ions to the cell surface of bacteria. The bacterial cell wall contains various functional groups, such as hydroxyl, amino, and sulfhydryl groups, which can interact with metal ions and facilitate their adsorption onto the cell surface. Biosorption is a rapid process and can contribute significantly to the overall metal accumulation by bacteria.

7.5.2. Metal Transformation: Bacteria possess enzymes that can transform heavy metals. They may use enzymatic reactions to alter the chemical form of metals, converting them into less toxic or more easily removable forms. For example, bacteria can oxidize metals or reduce them to less toxic states. These transformations can enhance the mobility or availability of metals, making them more amenable to removal or degradation. The detail process involves:

(i) *Reduction*: Bacteria can catalyze the reduction of certain metal ions, converting them from a higher oxidation state to a lower oxidation state. This reduction process can result in the precipitation of metals as less soluble and less toxic forms, facilitating their removal from the environment. For example, bacteria such as *Shewanella* and *Geobacter* are known to reduce toxic metals like uranium (U) and technetium (Tc) from their soluble forms to insoluble and less mobile precipitates.

(ii) *Oxidation*: Some bacteria have the ability to oxidize certain metals, converting them to higher oxidation states. This oxidation process can enhance the solubility and mobility of metals, making them more amenable to removal or extraction from contaminated sites. For instance, bacteria belonging to the Acidithiobacillus genus are capable of oxidizing sulfide minerals, leading to the release of bound metals like copper (Cu), zinc (Zn), and iron (Fe).

(iii) *Complexation:* Bacteria can produce extracellular compounds that form complexes with metal ions, leading to the formation of stable metal complexes. These metal complexes can alter the speciation and behavior of metals in the environment. For example, bacteria may produce siderophores, which are small organic molecules that bind to metal ions, particularly iron (Fe), with high affinity. Siderophore-mediated complexation can influence the bioavailability and mobility of metals.

(iv) *Methylation*: Certain bacteria are capable of methylating metal ions, particularly mercury (Hg), leading to the formation of methylated metal compounds. Methylmercury (MeHg) is a well-known example of a methylated mercury compound. Methylmercury is a highly toxic form of mercury that can bioaccumulate in the food chain. Bacteria such as sulfate-reducing bacteria and methanogenic bacteria play a role in methylating mercury.

(v) *Precipitation and Biomineralization*: Bacteria can induce the precipitation of metals as minerals or mineral-like compounds. They can promote the nucleation and growth of metal precipitates on their cell surfaces or in their extracellular environment. This process is known as biomineralization. Bacteria can influence the formation of various metal minerals, including sulfides, carbonates, and phosphates. The precipitation and biomineralization of metals can reduce their solubility and mobility.

7.5.3. Metal Sequestration: Bacteria can sequester heavy metals by binding them to specific proteins or chelating molecules. Metal-binding molecules, known as chelators, have high affinity for metal ions and can form stable complexes with them. This sequestration prevents the metals from interacting with cellular components or being released back into the environment. Some bacteria also form inclusion bodies or granules within their cells, where heavy metals can be stored. The detail process involves:

(i) *Biosorption*: Bacteria can accumulate metal ions on their cell surfaces through a process called biosorption. The cell surface may contain functional groups such as carboxyl, amino, and phosphate groups, which can bind to metal ions through electrostatic interactions or coordination. The metal ions become physically trapped on the bacterial cell surface, effectively removing them from the surrounding environment.

(ii) *Extracellular precipitation*: Some bacteria can produce extracellular substances, such as exopolysaccharides or metal-binding proteins, which can bind metal ions and form insoluble complexes. These complexes can precipitate out of solution, effectively immobilizing the metals and reducing their toxicity.

(iii) *Intracellular sequestration*: Bacteria can also sequester metals inside their cells. They may take up metal ions through transport proteins embedded in their cell membranes. Once inside the cell, the metals can be sequestered in various ways. For example, bacteria can store metals in specialized compartments called intracellular granules or vesicles. These granules may contain metal-binding proteins, metallothioneins, or metal sulfide complexes, which help to reduce the toxicity of the metals and prevent their interaction with cellular components.

(iv) *Redox transformation*: Some bacteria possess enzymes that can catalyze the redox transformation of metal ions. For instance, certain bacteria can reduce toxic forms of metals, such as hexavalent chromium (Cr(VI)) or arsenate (As(V)), to less toxic or insoluble forms, such as trivalent chromium (Cr(III)) or arsenite (As(III)). This transformation reduces the mobility and bioavailability of the metals, making them less harmful to the environment.

(v) *Chelation*: Bacteria can produce small organic molecules called chelators or siderophores, which have a high affinity for metal ions. These chelators can bind to metal ions and form stable complexes, preventing the metals from interacting with other molecules or precipitating out of solution. This sequestration mechanism enhances metal uptake by bacteria and helps to reduce the metal concentrations in the environment.

7.5.4. Redox Reactions: Certain bacteria have the ability to carry out redox reactions that can facilitate the remediation of heavy metals. For example, some bacteria are capable of reducing toxic metal ions to less toxic or insoluble forms, making them less mobile and less bioavailable. Conversely, other bacteria can oxidize metals, converting them to forms that are more amenable to removal or degradation. The detail process involves:

(i) *Reduction of metal ions*: Certain bacteria possess enzymes known as reductases that can facilitate the reduction of metal ions to less toxic or insoluble forms. For example, bacteria like Shewanella and Geobacter are capable of reducing highly toxic metals such as uranium, chromium, and technetium. These bacteria use their respiratory systems to transfer electrons from an electron donor (e.g., organic compounds) to metal ions, effectively reducing them to less mobile or less toxic states.

(ii) *Oxidation of metals*: Conversely, some bacteria are capable of oxidizing metals to alter their chemical speciation or enhance their removal. Acidophilic bacteria, such as Acidithiobacillus ferrooxidans, are known for their ability to oxidize sulfide minerals containing metals like iron, copper, and zinc. By oxidizing these minerals, bacteria can promote the release of metal ions from solid matrices, making them more accessible for subsequent remediation steps.

(iii) *Reductive dissolution*: Bacteria can facilitate the reductive dissolution of metal-bearing minerals, thereby releasing metal ions into solution. This process involves the reduction of specific mineral components by bacterial enzymes, resulting in the breakdown of the mineral structure and the subsequent release of metal ions. Bacteria like Acidithiobacillus spp. are capable of reductively dissolving iron-bearing minerals (e.g., iron oxides), liberating iron ions in the process.

(iv) *Methylation and demethylation*: Certain metals, such as mercury, can undergo methylation, a process in which bacteria add a methyl group (CH3) to the metal ion. This transformation can occur in anaerobic environments and is carried out by specific methylating bacteria. Methylation can convert less toxic forms of mercury (e.g., inorganic mercury) into highly toxic methylmercury, which has a high bioaccumulation potential. Conversely, other bacteria can facilitate the demethylation of methylmercury, converting it back to less toxic forms.

7.5.5. pH Adjustment: Bacteria can modify the pH of their surroundings through metabolic activities. This can influence the speciation and solubility of heavy metals. By altering the pH, bacteria can promote the precipitation or immobilization of metals, reducing their bioavailability and potential toxicity. The detail process involves:

(i) *pH-induced solubility changes*: The solubility of metals is highly dependent on the pH of the environment. Bacteria can modify the pH of their surroundings to promote the dissolution or precipitation of metal contaminants. For example, some bacteria can produce organic acids (e.g., citric acid, oxalic acid) as metabolic byproducts. These organic acids can lower the pH of the environment, enhancing the solubility of certain metals. By adjusting the pH, bacteria can increase the mobilization of metal ions, making them more accessible for subsequent bioremediation processes.

(ii) *pH-dependent metal complexation*: Bacteria can alter the speciation and bioavailability of metal ions through pH-dependent complexation reactions. Different metals exhibit different complexation behaviors at varying pH ranges. Bacteria can release specific chelating compounds or siderophores that bind to metal ions at particular pH levels. These complexes can either enhance or reduce the mobility and toxicity of metals. For instance, at higher pH levels, some bacteria produce siderophores that bind to metal ions, promoting their uptake by the bacterial cells and reducing their availability in the environment.

(iii) *pH regulation for bacterial metabolism*: Bacteria maintain their intracellular pH within a certain range for optimal metabolic activity. When exposed to metal-contaminated environments, bacteria may experience changes in the pH due to metal interactions. To mitigate the toxic effects of metals and maintain cellular function, bacteria employ pH adjustment mechanisms to regulate their internal pH. This pH regulation helps ensure the survival and metabolic activity of the bacteria during metal bioremediation processes.

(iv) *Control of bacterial growth and activity*: pH plays a critical role in bacterial growth and activity. Different bacteria have specific pH ranges at which they thrive. By adjusting the pH of their surroundings, bacteria can create conditions favorable for their growth and metabolic activity. This, in turn, influences their ability to carry out metal bioremediation processes effectively. Bacteria can produce alkaline or acidic metabolic byproducts to modulate the local pH, creating an environment that supports their growth and enhances their metal remediation capabilities.

7.5.6. Synergistic Interactions: Bacterial communities often consist of different species with complementary abilities. Some bacteria may specialize in metal uptake, while others excel in metal transformation or degradation. In these microbial consortia, synergistic interactions can occur, where one bacterium's activities support or enhance the actions of another. This cooperative behavior can improve overall metal bioremediation efficiency. The detail process involves:

(i) *Complementary metabolic capabilities*: Different bacterial species possess diverse metabolic capabilities, allowing them to utilize various energy sources and perform specific enzymatic reactions. In metal-contaminated environments, synergistic interactions occur when one bacterium produces metabolic byproducts or enzymes that can benefit another bacterium. For example, one bacterium may produce organic acids as byproducts of its metabolism, which can lower the pH and enhance the solubility of metals, benefiting another bacterium that can then utilize these solubilized metals for its own metabolic processes.

(ii) *Cooperative biofilm formation*: Bacteria often form biofilms, which are complex communities of microorganisms embedded in a matrix of extracellular polymeric substances (EPS). In metal bioremediation, different bacterial species within a biofilm can interact synergistically to enhance metal removal. Some bacteria may produce EPS components that bind to metals, promoting their immobilization and reducing their bioavailability. Other bacteria within the biofilm may contribute to the formation of channels or networks that facilitate the transport of metals through the biofilm, increasing the overall efficiency of metal uptake and sequestration.

(iii) *Redox interactions*: Bacteria with different redox capabilities can engage in synergistic interactions during metal bioremediation. For instance, some bacteria may oxidize certain metals, releasing them from solid matrices, while others with reducing capabilities can subsequently reduce these metals to less toxic or less mobile forms. This redox cycling between different bacterial species can enhance the transformation and detoxification of metals in the environment.

(iv) *Nutrient exchange and cooperation*: Bacteria in metal bioremediation processes can exhibit synergistic interactions through the exchange of nutrients and cooperative behaviors. For example, one bacterium may produce extracellular enzymes that break down complex organic compounds, releasing simpler forms that can be readily utilized by other bacteria in the community. This nutrient exchange and cooperation between bacteria can create an environment conducive to their growth and metabolic activities, ultimately contributing to the overall efficiency of metal bioremediation.

(v) *Indirect interactions through quorum sensing*: Bacteria can communicate and coordinate their activities through quorum sensing, a process by which they release and detect signaling molecules called autoinducers. In metal bioremediation, bacteria may engage in synergistic interactions mediated by quorum sensing. For example, certain bacteria may produce autoinducers in response to metal contamination, triggering the expression of genes involved in metal resistance or metal transformation in neighboring bacteria. This coordinated response can lead to enhanced metal removal and detoxification.

7. Detoxification Pathways: Bacteria possess detoxification mechanisms to protect themselves from the toxic effects of heavy metals. These mechanisms often involve the production of metal-detoxifying enzymes, such as metallothioneins, which can bind to heavy metals and prevent their harmful interactions with cellular components. The same detoxification pathways can be harnessed to aid in the bioremediation of heavy metal-contaminated environments. Here's how metallothioneins can contribute to metal bioremediation:

(i) Metal sequestration: Metallothioneins have a strong binding capacity for heavy metals. When introduced into metal-contaminated environments, they can bind to metal ions and form stable complexes, effectively sequestering the metals and reducing their bioavailability. This sequestration prevents the metals from interacting with other molecules or organisms, mitigating their toxic effects and reducing their potential for further environmental contamination.

(ii) Metal uptake enhancement: Metallothioneins can also enhance the uptake of heavy metals by bacteria or other organisms involved in metal bioremediation. By binding to metal ions, metallothioneins increase the concentration of metal species that are readily accessible for uptake. This can promote the efficiency of metal uptake by microorganisms, plants, or other bioremediation agents, enhancing their ability to remove metals from contaminated sites.

(iii) Protection against metal-induced oxidative stress: Heavy metals can induce oxidative stress in cells by generating reactive oxygen species (ROS). Metallothioneins possess antioxidant properties and can scavenge ROS, providing protection against metal-induced oxidative damage. By reducing oxidative stress, metallothioneins help maintain the cellular integrity and functionality of bioremediation agents, enabling them to better withstand metal exposure and carry out effective remediation processes.

(iv) Facilitation of metal detoxification pathways: Metallothioneins can interact with other detoxification enzymes and proteins, facilitating their metal detoxification pathways. For example, metallothioneins may interact with enzymes involved in metal reduction or oxidation reactions, aiding in the transformation of toxic metal species into less harmful forms. This cooperation between metallothioneins and other metal-detoxifying enzymes enhances the overall efficiency of metal bioremediation processes.

7.6. Role of Phytochelatins in Heavy Metal Detoxification

Phytochelatins (PCs) are small metal-binding peptides found in plants, particularly in the roots of plants. Phytochelatins (PCs) are synthesized in the cytosol in response to the heavy metal stress and act as the most efficient heavy metal chelators in the plant system. PCs react with the heavy metal ions to form PC-metal or PC-metalloid complexes, which are sequestered in the vacuoles (Goldsbrough, 2020). Phytochelatins were named cadystins when they were first discovered as Cd chelators in *Schizosaccharomyces pombe.* The principal classes of metal chelators include, metallothioneins, organic acids, and amino acids. Metallothioneins (MTs) are a family of small, cysteine-rich proteins found in a wide range of organisms, including humans, animals, plants, and microorganisms. These proteins play a crucial role in regulating and maintaining essential metal ions, particularly zinc and copper, within cells. Metallothioneins are often referred to as metal-binding proteins or metalloproteins due to their ability to bind and store metal ions. These compounds play a crucial role in the plant's defense mechanism against heavy metal toxicity, which can be harmful to their growth and development.

Plant root systems acquire micronutrients such as iron (Fe), zinc (Zn), copper (Cu), and others from their surrounding terrestrial or aquatic environments in order to grow. Non-essential metals such as cadmium (Cd), arsenic (As), lead (Pb), and mercury (Hg) may also be present at low to high levels in these supplies. Plants follow numerous techniques in response to heavy metal toxicity, which include immobilization, exclusion, chelation, and compartmentalization of the metal ions, and expression of the general stress responses. Several plants have been identified that exhibit the unusual aptitude to exist under toxic conditions in heavy metal contaminated locations and have also been proven to accumulate a large amount of such metals within their biomass (hyperaccumulators). Plants have a unique physiological transport of nutrients such as calcium, iron, magnesium, manganese, cobalt, and zinc. Cd competes with these critical nutrients for trafficking by occupying transmembrane nutrient transporters. Cd reaches the root via cortical tissue and is likely to accumulate in the roots. However, it enters the xylem via an apoplastic and/or symplastic pathway for further transport to shoots and may be complexed with a variety of ligands, including organic acids and/or PCs. Once in the root, Cd damages cells, particularly nucleoli, and inhibits various enzymatic functions, including nitrate reductase and ribonuclease activity. Cd leads to Fe (II) deficiency in shoot tissues of plants, and hence affects photosynthesis by damaging the light harvesting complex II and photosystems I and II and by increasing the non-photochemical quenching.

Cd has no known beneficial biological functions and is harmful to plant cells even at low concentrations. Cd concentrations of 5-10 μg per gram of dry mass in leaves are harmful to most plants. However, there have been reports of roots of the ecotypes of a few genera of plants that thrive in Cd-enriched soils. A list of Cd-hyperaccumulator plants and their ecotypes is presented. These plants also have a defensive mechanism that restricts the entry of Cd into the xylem and prevents metal accumulation in shoot tissues, mainly through the creation of Cd-chelators in the root zone. As previously stated, chelators like PC production can occur in cells exposed to numerous metals and metalloids such as Cu^{2+}, Zn^{2+}, As^{2+}, Pb^{2+}, and Ag^{+}. Cd concentrations in plant shoots vary greatly in nature, and this is due to environmental variables and phylogenetic diversity. Species in the Caryophyllales and Lamiales groups collect Cd in shoots at substantially higher levels than other species.

The most significant action in plant strategy is the selection and optimal uptake of heavy metals required for growth, as well as the rejection of metals that are not required. Stress caused by heavy metals in plants requires a sophisticated system of signal transduction, which is essentially a two-step process, with process activation beginning with heavy metal detection. Reduction in key nutrients reduces plant vitality and ability to cope with (metal) stress. The creation of stress-related proteins and signaling molecules occurs when the plant detects the presence of metals, which causes the explicit activation of metal-responsive genes to combat the stress. Therefore, the normal defensive mechanisms for detoxifying heavy metals in plants and other animals may involve the creation of metal-specific legends (chelation) and the subsequent compartmentalization of the ligand–metal complexes in the cells.

7.7. Biomining as Bioremediation of Metal

The wide phrase "biomining" refers to the biological extraction of particular metals from their ores using plants, fungi, or prokaryotes (phytoextraction, also called phytomining or biomining). These organisms release various organic compounds that chelate metals from their surroundings and return them to the cell, where they are usually utilized for electron coordination. The usage of metals by microbes in their cells was found in the middle of the 20th century. Both stable metals like iron, copper, zinc, and gold as well as unstable elements like uranium and thorium can be utilized by some microorganisms.It is possible to cultivate large chemostats of microorganisms to extract metals from their media. Many commercially viable metal compounds can then be produced from these culture vats. Compared to conventional mining, biomining is a more environmentally friendly method. While biomining merely releases any metabolites or gases secreted by the bacterium, mining releases numerous contaminants.Models of bioremediation can employ the same idea. Environments tainted by metals, oils, or other hazardous substances can be infected with bacteria. By consuming these harmful substances to produce energy within their cells, the bacteria can purify the surroundings (Newsome and Falagán, 2021). Bacteria are capable of mining for metals, cleaning up oil spills, purifying gold, and producing energy from radioactive elements.

7.7.1 History of biomining

When Kenneth Temple, PhD, found that *Acidithiobacillus ferrooxidans* thrives in conditions rich in iron, copper, and magnesium, the first known biomining system was described in 1951. A. ferrooxidans was introduced into media that contained 2,000–26,000 parts per million of ferrous iron in Temple's experiment. He found that in high iron concentrations, the bacteria were more motile and grew more quickly. The media became extremely acidic due to the byproducts of the bacterial development, but the microorganisms continued to flourish there.Microorganisms have systems for detecting and absorbing metals for utilization in the cell, as demonstrated by Kenneth Temple's experiment. Modern, sophisticated biomining technologies are developed as a result of this discovery. The process of using microorganisms to extract metals from their growing medium is known as biomining. These systems can be applied to biohydrometallurgy, bioremediation, or even the commercial extraction of metals from ores. Later on, it was found that certain fungus also absorb metals from their surroundings.Certain bacteria have been demonstrated to have a mechanism for absorbing radioactive elements including thorium and uranium (Qiu et al., 2022).

7.7.2. Types of Biomining

» Stirred Tank Biomining

» Bioheaps

» In- situ Bioleaching

i. Stirred Tank Biomining

Leaching from substrates with high mineral concentrations is accomplished using this technique. For this purpose, specific kinds of stirred tank bioreactors that are insulated with cooling pipes or cooling jackets and lined with rubber or corrosion-resistant steel are employed. A common bacterium is *Thiobacillus*. Because the bio reactor is aerobic, aerators, pumps, and blowers offer a plentiful supply of oxygen throughout the process. This is a multi-phase process made up of numerous interconnected bio-reactors. In order to recover the mineral, the substrate is moved from one reactor to another before being cleaned with water and subjected to various chemical treatments. The method necessitates building big aerated tanks, which are typically grouped in a series, hence the term is quite self-explanatory. in order for one tank's discharge to be used as a raw material for another. This allows the reactor to function in continuous flow mode, adding new ore to the first tank while removing and treating the runoff from the last tank. To guarantee that the solids stay suspended in the liquid medium, the ore being treated is often crushed to a very fine particle size. To maintain the highest possible microbial density, mineral fertilizers in the form of KH_2PO_4 and $(NH_4)_2SO_4$ are also fed to the tanks.

Due to extremely high cost of stirred tank reactors, they are only used for highly valuable minerals or materials. For gold extraction for example, this technique is usually used when the ore body contains high concentration of arsenopyrite (AsFeS).

ii. Bioheaps

Bioheaps are large amount of low grade ore and effluents from extraction processes that contain trace amounts of minerals. Such effluents are usually stacked in the large open space heaps and treated with microorganisms to extract the minerals. Bioheaps are also called biopiles, biomounds and biocells. In this technique, the ore is arranged in large heaps. During heap leaching, an aqueous mixture of microorganisms is sprinkled over the leach pile. Then, the solution is collected and processed to help recover even more metal. They are also used for biodegradation of petroleum and chemical wastes. The low grade ore are crushed and acid treated ore is then agglomerated so that the finer particles get attached to the coarser ones and then treated with the water and effluent liquid. This is done to optimize moisture content in the ore bacteria that is inoculated along with the liquid. The ore is then stacked in the large heaps of 2-10m feet high with aerating tube to provide air supply to the bacteria thus promoting bio oxidation.

iii. In- situ Bioleaching

Instead of gathering the ore and transporting it to an extraction facility far from the mine, this approach extracts the mineral straight from the mine. After a traditional extraction procedure is finished, in-situ biomining is typically used to extract trace amounts of minerals from the ores. The mine is treated with water and acid solution containing bacterial inoculums after being blasted to decrease the size of the ore and boost permeability. Pipes or shafts are used to provide air. Growing bacteria cause bio-oxidation, which occurs in-situ and extracts mineral from the ore.

7.7.3. Process of recovery of metals through biomining

Minerals are recovered from ores by the microorganisms mainly by two mechanisms: (a) Oxidation and (b) Reduction.

a. Oxidation

The microorganisms like *Thiobacillus ferroxidans* and *T. thioxidans* are used to release iron and sulfur respectively. *T. ferroxidans* oxidize ferrous ion to ferric ion.

$$4Fe + O_2 + 4H = Fe + 2H_2O$$

b. Reduction

Bacteria like *Desulfovibro desulfuricans* play an active role in reduction of sulfates which results in the formation of hydrogen sulfides.

$$4H_2 + H_2SO_4 \rightarrow H_2S + 4H_2O$$

7.7.4. Biological tools use for biomining

A variety of mineral oxidizing bacteria readily found can easily oxidize iron and sulfur containing minerals. These include the iron- and sulfuroxidizing *Acidithiobacillus ferrooxidans* (previously, *Thiobacillus ferrooxidans*), the sulfuroxidizing *Acidithiobacillus thiooxidans* (previously *Thiobacillus thiooxidans*) and *Acidithiobacillus caldus* (previously, *Thiobacillus caldus*) and the ironoxidizing *Leptospirillum ferrooxidans* and *Leptospirillum ferriphilum.*

Several species of fungi can be used for biomining. Fungal strains *Aspergillus niger* and Penicillium simplicissimum were able to mobilize Cu, Sn, Al, Ni, Pb and Zn. Similarly, 'Phytomining' is based on the tendency of some plant species to bioaccumulate excessive amounts of metals from their host rock. The plants, called hyperaccumulators are grown on highly mineralized soils or post-mine lands and their yield (bio-ore) is used as a pure metal source. Compared to the bacterial mining, these technologies are not so popular primarily because of the longevity of these processes and so their unprofitability (Li and Wen, 2021).

7.7.5. Role of biomining in bioremediation

Biomining and bioremediation are two distinct processes with different primary goals, but there can be some overlap in their applications when it comes to environmental remediation. Here's how biomining can play a role in bioremediation:

i. **Metal Removal and Recovery**: Biomining is the process of using microorganisms to extract valuable metals from ores or mine tailings. In some cases, biomining can be used to extract and recover metals that have contaminated the environment. These microorganisms are capable of selectively leaching metals from contaminated soil or water, effectively removing them from the environment. This metal recovery process can contribute to the remediation of metal-contaminated sites.

ii. **Stabilization of Metal-Contaminated Sites**: Biomining can also play a role in stabilizing metal-contaminated sites. By extracting metals from the contaminated soil or water, biomining can reduce the bioavailability of these metals, making them less toxic to plants and organisms. This can help in the remediation and restoration of ecosystems affected by metal pollution.

iii. **Reduction of Acid Mine Drainage:** Acid mine drainage (AMD) is a significant environmental issue associated with mining activities, particularly in abandoned mines. Biomining processes can help reduce AMD by promoting the growth of acidophilic microorganisms that can neutralize acidic mine waters and precipitate metal sulfides, preventing the release of toxic metals into surrounding water bodies.

iv. **Rehabilitation of Mining Sites**: After biomining has been used to extract valuable metals from ores, the remaining materials, such as mine tailings, may still contain residual contaminants. Bioremediation techniques can be employed to further clean up these materials by using microorganisms to break down or transform contaminants, improving the overall environmental quality of the site.

v. **Integration with Phytoremediation**: Phytoremediation is a process that uses plants to remove, stabilize, or degrade contaminants from soil or water. Biomining can complement phytoremediation by removing metal contaminants from the environment, which can make it easier for plants to thrive and remediate other pollutants.

While biomining primarily focuses on metal extraction and recovery, it can have beneficial side effects in terms of bioremediation by reducing metal contamination and mitigating associated environmental issues. However, it's important to note that the primary purpose of biomining is metal recovery, and any bioremediation benefits are typically secondary to this main objective. Bioremediation strategies can be more targeted and versatile when the primary goal is environmental cleanup and the removal of various contaminants beyond metals.

7.7.6. Microbes useful for biomining operations

On Earth, there are many different kinds of microorganisms with different capacities. As a result, it's critical to pinpoint the exact kinds that can carry out biooxidation and bioleaching efficiently (Wong et al., 2020).

The chemophilic, somewhat thermophilic bacterium *Thiobacillus ferrooxidans* may generate energy by oxidizing inorganic substances like iron and sulfur. It is the bacteria that is most frequently employed in biomining. Other bacteria that are frequently used to extract different minerals include *Sulfolobus acidocaldarius, Thermothrix thiopara, T. thioxidans*, and *S. brierleyi.*

Sulfur is extracted using Thermothrix thiopara, a highly thermophilic bacterium that can withstand temperatures between 60 and 75 degrees Celsius. In order to boost the rate of biooxidation and, consequently, the mineral yield through biomining, methods such as genetic engineering and conjugation are employed to create bacteria with desired traits. Identification of biomining bacteria found in colonies of other bacteria is also crucial.

7.7.7. Characteristics of bacteria used in biomining

Bacteria are most suitable microorganism for the extraction of minerals from low grade ores. The characteristics of the bacteria used in bio-mining are:

1. Bacteria which are thermophilic that is bacteria which can survive at high temperature are selected for biomining, as mineral extraction involves the high temperature procedures.
2. Bacteria which are chemophilic, as biomining uses both strong acid and alkalis.
3. Autotropic bacteria are selected for biomining. (Bacteria, which has the ability to produce energy from inorganic compounds through photosynthesis or chemosysnthesis).
4. Bacteria which have the capacity to form biofilms are selected for biomining purpose.

Important bacteria involving in biomining:

Acidithiobacillus ferrooxidans, At. Ferrooxidans, Acidimicrobium ferrooxidans., Ferroplasma acidiphilum, Ferroplasma acidarmanus, Sulfolobus metallicus, S. metallicus, Acidianus brierleyi, Acidianus infernus, Ad. Ambivalens, Bacillus licheniformis

Biomining - Environment Friendly

The best method of sustaining the environment is such that returns back all the components (wastes) in a recyclable way so that the waste becomes useful and helps the biotic and abiotic relationship to maintain an aesthetic and healthy equilibrium that characterizes an ideal environment. The Biomining method is cheap, reliable, efficient and also environment friendly method of mineral recovery. Using biotechnology, efficiency of biomining can be increased by using genetically modified microorganisms. Biomining contributes to sustainable development in the same way as all microorganism-mediated processes do. It uses existing organisms and mechanisms in nature. Traditional mining is an especially toxic process involving the use of chemicals like cyanide. Although the process of biomining does not yet completely eliminate the use of harmful chemicals, it allows for a lessened use, resulting in lower production costs of cleaning up the mining processes.

7.7.8. Future of biomining

Biomining or the "mining of the future". Indeed, it is much cheaper and greener than traditional mining - there are a lot fewer CO2 emissions and carbon and water foot prints are lower than using conventional technology.

The future of biomining is challenging, as it offers advantages of operational simplicity, low capital and operating cost and shorter construction times that no other alternative process can provide. In addition, minimum environmental impact and the use of this technology in the mining industry are set to increase. Once commercial scale high-temperature processes have been designed, the variety of minerals that will become acquiescent to biomining will increase. Increased concern regarding the effect of mining on the environment is likely to improve the competitive advantage of microbially based metal recovery processes.

Chapter - 8

Biodegradation of Xenobiotics Compound

8.1. Introduction

Xenobiotic refers to any foreign substance or compound that is not naturally produced or expected to be found within an organism's biological system. The term "xeno" originates from the Greek word meaning "foreign" or "strange." Xenobiotics can include a wide range of chemicals, such as drugs, environmental pollutants, pesticides, industrial chemicals, and even certain food additives. When xenobiotics enter an organism's body, they may undergo various processes, including absorption, distribution, metabolism, and elimination. The body's response to xenobiotics can vary depending on the specific compound and the organism's ability to metabolize or detoxify it. Some xenobiotics are harmless and may be eliminated without causing any adverse effects, while others can have toxic or harmful effects on the organism. In many cases, organisms have developed specific mechanisms to metabolize and eliminate xenobiotics. These mechanisms often involve enzymes in the liver or other organs, which modify the chemical structure of the xenobiotic to make it more water-soluble and easier to eliminate through urine or feces. However, some xenobiotics can overwhelm these detoxification mechanisms, leading to toxicity and adverse health effects. The study of xenobiotics and their effects on living organisms is an important field within toxicology and environmental science. Scientists aim to understand how xenobiotics interact with biological systems, assess their potential risks to human health and the environment, and develop strategies to mitigate or minimize their harmful effects (Hashmi et al., 2017). An example of a group of enzymes involved in xenobiotic metabolism is hepatic microsomal cytochrome P450. These enzymes that metabolize xenobiotics are very important for the pharmaceutical industry because they are responsible for the breakdown of medications. A species with this unique cytochrome P450 system is *Drosophila mettleri*, which uses xenobiotic resistance to exploit a wider nesting

DOI: 10.1201/9781003787549-8

range including both soil moistened with necrotic exudates and necrotic plots themselves.

In general, xenobiotic chemicals fall into two categories. They could be nondegradable (recalcitrant) or biodegradable. While recalcitrant chemicals are resistant to degradation by any process, biodegradable xenobiotic substances are broken down by bacteria or other reactions. Halocarbons, polychlorinated biphenyls, oil mixes, synthetic polymers, alkyl benzyl sulphonates, and other substances can be classified as resistant xenobiotics. Water does not dissolve synthetic polymers like nylon and plastics. Many of the chemicals in oil are biodegradable and degrade at varying rates, making it a pollutant as well. Oil is refractory mostly because it is insoluble in water and because some of its components are poisonous.

Important features of xenobiotics compounds:

i. Xenobiotics are highly stable in nature.

ii. They are not recognized as substrates by degenerative organisms.

iii. Their solubility in water is very low or insoluble.

iv. They are highly toxic i.e toxicity of the compounds which are produced is very high.

v. Since xenobiotic compounds' molecular weight is very high they reduce the chances of entry into microbial cells.

vi. These compounds have branched linear chains which are resistant to the biodegradation process and contain halanongs present in place of hydrogens.

vii. There are other compounds present including nitrates, amino, etc.

viii. The chemicals' inability to promote the creation of degrading enzymes

ix. Xenobiotic Compounds are mostly recalcitrant i.e they don't cooperate with other chemicals.

2.1. Types of xenobiotic compounds

Compounds types	Features
Halocarbons	» Different numbers of halogen (e.g., CI, Br, F (fluorine), I) atoms replace H atoms in these compounds. » They're employed as solvents (chloroform, CHCL3), propellants in cosmetics, paints, and other spray cans, in cooling system condenser units (Freons, CCI3F, CCl2F2, CClF3, CF4), and as pesticides (DDT, BHC, lindane, and so on) (dalapon, 2, 4-D, 2, 4, 5-T etc.). » Chloroform, freons, and other C1-C2 haloalkanes are volatile and escape into the atmosphere, where they deplete the protective ozone (O3) layer, resulting in increased UV radiation. » Pesticides (herbicides, fungicides, and insecticides) are sprayed on crops and then leak into water bodies, with many of them being biomagnified.
Polychlorinated biphenyls (PCBs)	» Two covalently connected benzene rings with halogens replacing for H make up these molecules. PCBs are utilized as plasticizers, transformer insulator coolants, and heat exchange fluids. » They are physiologically and chemically inert to varying degrees, with the number of chlorine atoms in the molecule increasing. » The following two categories of chemicals are refractory because of their halogenation as well as their cyclic structure (PCBs).
Synthetic polymers	» Plastics, such as polyethylene, polystyrene, polyvinyl chloride, and nylons, are made from these compounds and are utilized in clothes, wrapping materials, and other applications. » They are resistant to water because of their insolubility and molecular size.
Alkyl benzene sulphonates	» The sulphonate (— SO3–) group at one end resists microbial destruction, but the non-polar alkyl end (if branched) becomes recalcitrant (resistance increases with the degree of branching). » Currently, non-branched alkylbenzyl sulfonates are employed, which are biodegraded via -oxidation from their alkyl ends.

Oil mixtures	» Oil is a natural product that has numerous components and is biodegradable, with the various components degrading at varying rates. » Small oil seepages can be handled through biodegradation. » However, when significant leaks occur, pollution becomes a serious issue. » Oil is difficult to work with due to its insolubility in water and the toxicity of some of its constituents.
Other compounds	» A number of pesticides are based on aliphatic, cyclic ring structures with nitro-, sulfonate, methoxy-, amino-, and carbonyl group substitutions, as well as halogens. » They become recalcitrant as a result of these substitutions.

There are numerous examples of xenobiotics. Here are a few common examples:

» **Pharmaceuticals**: Medications and drugs are xenobiotics. When you take a prescription or over-the-counter drug, it's a foreign substance introduced into your body to treat or manage a particular condition.

» **Environmental pollutants**: Substances such as pesticides, herbicides, heavy metals, industrial chemicals, and air pollutants can be considered xenobiotics. These compounds are often released into the environment through human activities and can have harmful effects on ecosystems and human health.

» **Food additives**: Certain food additives, such as artificial sweeteners, artificial flavors, and preservatives, are xenobiotics. These substances are added to food products to enhance taste, appearance, or shelf life.

» **Personal care products**: Many ingredients in personal care products like shampoos, soaps, lotions, and cosmetics are xenobiotics. These products contain chemicals that are not naturally produced by the body and may have both desired effects and potential risks.

» **Household chemicals:** Various chemicals found in household cleaning products, detergents, and solvents can be classified as xenobiotics. These substances are designed to perform specific functions but may pose health risks if not used properly.

» **Recreational drugs**: Substances like alcohol, nicotine, marijuana, cocaine, and other illicit drugs are considered xenobiotics. These substances alter the normal functioning of the body and can have profound effects on physical and mental health.

These examples illustrate the diverse range of xenobiotics encountered in daily life, highlighting the importance of understanding their potential impacts on human health and the environment.

In the industrial revolution and urbanization era, the global environment's poisoning by a complex mixture of xenobiotics has become a major environmental threat worldwide.Because of their long-term persistence and slow or nonexistent biodegradation in ecosystems, xenobiotic contaminants like azodyes, phenolics, polycyclic aromatic hydrocarbons (PAHs), halogenated compounds, personal care products (PCPs), pharmaceutical active compounds (PhACs), pesticides, nitroaromatic compounds, triazines, and chlorinated compounds have a negative impact on the environment. After being released into the environment, xenobiotics infiltrate the food chain and have detrimental effects on both human and animal health at every trophic level. The public became interested in the bioaccumulation of xenobiotic compounds in the 1960s after DDT (dichlorodiphenyl-trichloroethane) and methyl mercury deposits were found in fish and wildlife. These contaminants also have toxic, mutagenic, carcinogenic, and teratogenic impacts on all living things. Thus, it is essential to eliminate hazardous, non-biodegradable xenobiotics from the environment. These xenobiotic compounds have been degraded by physical and chemical methods such as coagulation, filtration, adsorption, chemical precipitation, electrolysis, and ozonation.

8.2. Xenobiotics in the Environment

Xenobiotics are foreign substances that are not naturally found in the environment. They include various chemical compounds such as pesticides, pharmaceuticals, industrial chemicals, and personal care products. These substances can enter the environment through various routes, including industrial discharges, agricultural activities, improper waste disposal, and accidental spills. Once xenobiotics are released into the environment, they can have a range of effects on ecosystems and organisms. Some xenobiotics are biodegradable and can be broken down by natural processes over time. However, many xenobiotics are persistent and can accumulate in the environment, leading to long-term exposure and potential harm. Xenobiotics can have detrimental effects on both wildlife and human health. In ecosystems, they can disrupt the balance of populations, harm aquatic organisms, and impact the food chain (Croom, 2012). For example, certain pesticides have been linked to declines in bee populations, while pharmaceutical residues in water bodies can affect fish and other aquatic organisms. Human exposure to xenobiotics can occur through various pathways, including consumption of contaminated food and water, inhalation of polluted air, and direct contact with contaminated soil or water. Some xenobiotics have been associated with adverse health effects in humans, including developmental abnormalities, hormonal disruptions, cancer, and organ damage. Efforts are made to monitor and regulate the presence of xenobiotics in the environment. Regulatory agencies establish guidelines and standards for acceptable levels of various substances, and industries are encouraged to develop environmentally friendly alternatives. Additionally, wastewater treatment plants and other pollution control measures help to mitigate the release of xenobiotics into the environment. Research and technological advancements are also focused on developing better methods for detecting and removing xenobiotics from the environment. This includes the use of advanced analytical techniques, such as high-performance liquid chromatography and mass spectrometry, to identify and quantify xenobiotic compounds. Furthermore, bioremediation techniques, such as the use of specialized microorganisms, are being explored as a means to degrade or remove xenobiotics from contaminated sites.

8.3. Xenobiotic Pollution and Its Impact on the Environment

Environmental xenobiotic contamination is a worldwide issue brought on by human activities like population growth and urbanization. Widespread ecosystem contamination is caused by the massive quantity of dangerous substances discharged into the environment. Soil, sediment, and water include significant pollutants such pesticides, fertilizers, heavy metal ions, polycyclic aromatic hydrocarbons (PAHs), and oil derivatives.

Scientific and technological developments throughout the Industrial Revolution led to overuse of resources by humans, which devastated numerous ecosystems [19]. Another well-known cause of environmental pollution is the illogical use of pharmaceutical waste, veterinary medications, and human medications. Medicines have the ability to affect aquatic flora and wildlife more than other chemical molecules. Pharmaceuticals, however, are thought to pose very little risk of acute environmental damage. Although the situation may be different for long-term impacts, there is a significant lack of data regarding long-term effects and their toxicity (Miglani et al., 2022). Furthermore, despite the fact that many aquatic animals are exposed to toxins throughout their lives, there is little to no evidence of multi-generational life cycle impacts.

8.3.1. Impact of Xenobiotic on soil

Xenobiotics can contaminate soil through various means, including the use of pesticides, herbicides, industrial pollutants, and improper waste disposal. These chemicals can persist in the soil for extended periods, altering the soil's chemical and physical properties. Xenobiotics can contaminate soil through various means, including the use of pesticides, herbicides, industrial pollutants, and improper waste disposal. These chemicals can persist in the soil for extended periods, altering the soil's chemical and physical properties. Xenobiotics can disrupt the diversity and activity of soil microbial communities. Soil microorganisms play crucial roles in nutrient cycling, organic matter decomposition, and maintaining soil structure. Exposure to xenobiotics can lead to a decrease in microbial diversity and alter microbial community composition, which can have cascading effects on soil health and ecosystem functioning. Moreover, xenobiotics can interfere with the decomposition of organic matter in the soil. Organic matter provides essential nutrients and acts as a carbon source for soil microorganisms. When xenobiotics inhibit microbial activity, the breakdown of organic matter can be hindered, resulting in the accumulation of organic residues and reduced soil fertility. Xenobiotics that persist in the soil can potentially leach into groundwater, particularly if they are water-soluble. This can lead to the contamination of drinking water sources and pose a risk to human health. Some xenobiotics are highly persistent and can remain in the soil for years or even decades. Their long-term presence in soil can continue to exert negative effects on soil health, plant growth, and ecosystem functioning. This persistence can pose challenges for remediation efforts and may require specialized techniques to mitigate their impact (Jayaramaiah et al., 2022).

8.3.2. Impact of Xenobiotic on water

Xenobiotics can have significant impacts on water bodies when they are released into the environment. Xenobiotics can contaminate water bodies through various sources, such as industrial discharges, agricultural runoff, and improper disposal of chemicals. These substances can enter surface water bodies, such as rivers, lakes, and streams, as well as groundwater. Contamination can lead to elevated concentrations of xenobiotics in water, posing risks to aquatic organisms and potentially impacting the quality of drinking water sources. Xenobiotics can be toxic to aquatic organisms, including fish, invertebrates, and algae. Some xenobiotics have acute toxic effects, leading to immediate harm or death of aquatic organisms upon exposure. Chronic exposure to sub-lethal levels of xenobiotics can also have long-term effects, such as impaired growth, reproductive abnormalities, and changes in behavior. The overall biodiversity and ecological balance of aquatic ecosystems can be disrupted by xenobiotic pollution. Xenobiotics can bioaccumulate in aquatic organisms. This means that they can accumulate in the tissues of organisms over time, particularly in fatty tissues. As organisms consume contaminated food or water, xenobiotics can biomagnify, meaning their concentrations increase along the food chain. This can lead to high levels of xenobiotics in top predators, such as fish and birds, which can have significant health impacts on these species and potentially affect human health through the consumption of contaminated seafood. Xenobiotics can disrupt important ecosystem processes in water bodies. For example, certain chemicals can inhibit photosynthesis in aquatic plants, leading to reduced oxygen production and potential oxygen depletion in the water. This can harm fish and other organisms that rely on oxygen for survival. Additionally, xenobiotics can interfere with nutrient cycling and other biochemical processes, affecting the overall functioning and health of aquatic ecosystems. Some xenobiotics are highly persistent in water, meaning they do not readily degrade or break down. These persistent xenobiotics can remain in water bodies for extended periods, continuously exerting their effects on aquatic organisms and ecosystems. This persistence poses challenges for water quality management and may require specialized strategies for remediation and removal (Piwowarska and Kiedrzyńska, 2022).

8.3.3. Impact of Xenobiotic on plants

Xenobiotics can have various impacts on plants when they are present in the environment. Xenobiotics can inhibit plant growth and development. They may interfere with essential physiological processes, such as photosynthesis, respiration, and nutrient uptake. This can lead to reduced plant biomass, stunted growth, and impaired reproductive success. Some xenobiotics can cause physiological disorders in plants. For example, certain herbicides can disrupt hormonal balance in plants, leading to abnormal growth patterns, leaf curling, and chlorosis (yellowing of leaves). These disorders can impact plant health and reduce crop productivity. Xenobiotics can cause direct damage to plant tissues. Certain chemicals, such as air pollutants or phytotoxic industrial waste, can result in leaf necrosis, tissue browning, or wilting. Pesticides and herbicides can also lead to damage or death of non-target plant species. Xenobiotics that enter the soil can impact plants indirectly. Contaminated soil can hinder nutrient availability, impair root development, and affect water uptake by plants. Additionally, some xenobiotics may be taken up by plants and accumulate in their tissues, potentially leading to toxicity and reduced plant performance. Xenobiotics can induce genetic and molecular changes in plants. They may cause DNA damage, mutations, or alterations in gene expression. These changes can affect plant growth, development, and stress response mechanisms. Xenobiotics can have broader ecological impacts by affecting plant populations and interactions with other organisms. For example, herbicides may unintentionally eliminate non-target plant species, disrupting ecological communities and reducing biodiversity. Changes in plant populations can also impact pollinators, herbivores, and other dependent organisms (Singh et al., 2016).

8.3.4. Impact of Xenobiotic on marine life

Xenobiotics can be toxic to marine organisms, leading to adverse effects on their health and survival. Different xenobiotics have varying degrees of toxicity, and their effects depend on factors such as concentration, exposure duration, and the sensitivity of the species. Marine organisms can experience acute toxicity, resulting in immediate death or illness, as well as chronic effects, including reproductive abnormalities, impaired growth, and behavioral changes. Xenobiotics have the potential to bioaccumulate in the tissues of marine organisms. This means that these substances can accumulate in higher concentrations within the bodies of organisms compared to the surrounding environment. Through biomagnification, xenobiotics can increase in concentration as they move up the food chain, resulting in higher levels of contamination in top predators. This can pose significant risks to marine mammals, large predatory fish, and seabirds. Xenobiotics can interfere with reproductive processes and development in marine organisms. Some xenobiotics can disrupt hormonal systems, affecting reproductive organs, fertility, and sexual development. This can lead to reduced reproductive success, population declines, and impaired growth and survival of offspring. Xenobiotic pollution can cause damage to marine ecosystems. For example, certain xenobiotics can harm coral reefs, leading to coral bleaching, reduced calcification, and decreased resilience. Coral reefs are critical habitats that support a wide range of marine species. Damage to coral reefs can disrupt the entire ecosystem, leading to a loss of biodiversity and ecological imbalances. Xenobiotics can affect the structure and functioning of marine food webs. Contamination of primary producers, such as phytoplankton and algae, can have cascading effects on higher trophic levels. Changes in the abundance or quality of food sources can impact the growth, reproduction, and survival of organisms throughout the food web. Xenobiotics can reduce the resilience of marine ecosystems to natural and human-induced stressors. The presence of xenobiotics can weaken the ability of organisms to cope with environmental changes, such as temperature fluctuations, pollution events, or disease outbreaks. This can make marine ecosystems more vulnerable to further disturbances and disrupt their ability to recover.

8.3.5. Impact of Xenobiotic on terrestrial animals

Xenobiotics can be toxic to terrestrial animals, leading to adverse effects on their health and survival. Different xenobiotics have varying degrees of toxicity, and their effects depend on factors such as the concentration, exposure duration, and the sensitivity of the species. Terrestrial animals can experience acute toxicity, resulting in immediate death or illness, as well as chronic effects, including reproductive abnormalities, impaired growth, and behavioral changes. Xenobiotics can bioaccumulate in the tissues of terrestrial animals. This means that these substances can accumulate in higher concentrations within the bodies of animals compared to the surrounding environment. Through biomagnification, xenobiotics can increase in concentration as they move up the food chain, resulting in higher levels of contamination in top predators. This can pose significant risks to carnivorous animals and animals that consume contaminated food sources. Xenobiotics can interfere with the reproductive processes and development of terrestrial animals. Some xenobiotics can disrupt hormonal systems, affecting reproductive organs, fertility , and sexual development. This can lead to reduced reproductive success, population declines, and impaired growth and survival of offspring. Xenobiotics can impact the behavior and physiology of terrestrial animals. Exposure to certain chemicals can alter feeding behavior, migration patterns, and the ability to find mates. Xenobiotics can also affect immune function, metabolism, and other physiological processes, potentially making animals more susceptible to diseases or reducing their overall fitness. Xenobiotics can indirectly contribute to habitat loss and fragmentation, which can have detrimental effects on terrestrial animals. Pollution caused by xenobiotics can contaminate soil, water, and air, leading to degradation of habitats. Additionally, industrial activities associated with the production and use of xenobiotics can result in deforestation, habitat destruction, and the loss of critical wildlife habitats. Xenobiotics can affect keystone species, which play critical roles in maintaining the structure and function of ecosystems. For example, exposure to xenobiotics can harm pollinators such as bees and butterflies, which are essential for the pollination of many plant species. Disruption of pollinator populations can have cascading effects on plant reproduction and ecosystem dynamics.

8.3.6. Impact of Xenobiotic on human health

Xenobiotics can be toxic to humans, leading to adverse health effects. Exposure to certain chemicals, such as heavy metals, pesticides, or industrial pollutants, can cause acute toxicity, resulting in immediate symptoms or illness. Chronic exposure to lower levels of xenobiotics can also have long-term health effects, including organ damage, developmental disorders, and an increased risk of chronic diseases such as cancer, cardiovascular diseases, or neurological disorders. Some xenobiotics can trigger allergic reactions in susceptible individuals. Certain chemicals, such as certain drugs, cosmetics, or food additives, can cause allergic responses ranging from mild skin rashes to severe anaphylactic reactions. Allergic reactions can vary widely depending on the individual's sensitivity and the specific xenobiotic involved. Xenobiotics, known as endocrine disruptors, can interfere with the normal functioning of the endocrine system in humans. These chemicals can mimic or block the action of hormones, leading to hormonal imbalances and potential health effects. Endocrine disruptors have been linked to reproductive disorders, developmental abnormalities, impaired fertility, and hormonal-related cancers. Some xenobiotics have carcinogenic properties, meaning they can increase the risk of developing cancer. Exposure to certain chemicals, such as asbestos, benzene, or certain pesticides, has been associated with various types of cancer, including lung, bladder, liver, and breast cancer. The effects of xenobiotics on cancer development can depend on the duration and intensity of exposure, individual susceptibility, and the specific xenobiotic involved. Certain xenobiotics can have neurotoxic effects, meaning they can damage the nervous system. Exposure to neurotoxic substances, such as lead, mercury, or certain pesticides, can lead to neurological disorders, cognitive impairments, developmental delays in children, and other nervous system dysfunctions. Xenobiotics can affect human reproductive health and development. Some chemicals, such as certain pesticides, industrial pollutants, or pharmaceuticals, can interfere with reproductive functions, fertility, and fetal development. They may increase the risk of infertility, miscarriages, birth defects, or developmental disorders in children.

8.4. Microbial Remediation of Xenobiotic Compounds

Microorganisms has the capability of degrading all naturally occurring compounds; this is known as the principle of microbial perfectness proposed by Alexander in 1965. Microorganism area also able to degrade many of the xenobiotic compounds, but they are unable to degrade many others (Bharadwaj, 2018). The compounds that resist biodegradation and thereby persist in the environment are called recalcitrant due to one or more of the following reasons:

- They are not recognized as substrate by the existing degradative enzymes,
- They are highly stable, i.e., chemically and biologically inert due to the presence of substitution groups like halogens, nitro-, sulphonate, amino-, methoxy- and carbamyl groups,

» They are insoluble in water, or are adsorbed to external matrices like soil,

» They are highly toxic or give rise to toxic products due to microbial activity,

» Their large molecular size prevents entry into microbial cells,

» Inability of the compounds to induce the synthesis of degrading enzymes.

» Lack of the perm-ease needed for their transport into the microbial cells.

Two different microbes can together be used to degrade xenobiotic compounds completely.

Use of mix cultures of microorganisms produce different enzymes that act on recalcitrant compounds and degrade them to simpler form. The microorganism that contain variety of mono and di-oxygenase enzymes are capable of degrading xenobiotics compounds into smaller compounds that are again taken up by other series of microbes and degraded wholly.

Depending on their fate in the air, water, soil, or sediment, xenobiotic compounds may become accessible to microorganisms in different environmental compartments where they are further transformed and degraded.

8.4.1. Microbial degradation of DDT

Microbial degradation of DDT residues is one method for removing DDT from soil. DDT residues can biodegrade in soil, albeit slowly. Several solutions are presented to improve in-situ deterioration. They include the introduction of DDT-metabolising microorganisms into contaminated soils and/or the manipulation of environmental circumstances to increase the activity of these bacteria. Ligninolytic fungus and chlorobiphenyl-degrading bacteria are interesting possibilities for cleanup. Flooding the soil and adding organic materials can improve DDT breakdown. Because biodegradation may be hampered by the microbe's lack of access to the contaminant, the soil may need to be pretreated with a surfactant.

Both DDD and DDE are DDT transition products. DDD is generated from DDT via reductive dechlorination, which is either microbially mediated or the product of chemical processes mediated in part by biomolecules. DDE is generated from DDT through photochemical processes in the presence of sunshine, as well as dehydrochlorination by bacteria and animals.

The presence of chlorinated substituents on these compounds has been linked to their persistence. DDT, for example, is resistive, although its non-chlorinated counterpart, diphenylmethane, is easily biodegradable.

DDTr are lipophilic, therefore they tend to accumulate in the fatty tissues of eating species along the food chain. Although there are no documented acute effects of ambient DDT or its metabolites on human health, the possibility of chronic consequences cannot be discounted. Some DDTr exhibit estrogenic activity, whereas DDE has recently been found as a potent androgen receptor. DDTr are lipophilic, therefore they tend to accumulate in the fatty tissues of eating species along the food chain. Although there are no documented acute effects of ambient DDT or its metabolites on human health, the possibility of chronic consequences cannot be discounted. Some DDTr exhibit estrogenic activity, whereas DDE has recently been found as a potent androgen receptor. DDTr are accumulated by people through consumption of agricultural products grown on contaminated soils. DDTr in soil are of concern as their uptake can lead to accumulation in primary products. Their removal from soil is therefore a priority.

DDTr biodegradation is mostly dependent on co-metabolism, in which bacteria growing at the expense of a growth substrate can convert DDTr without gaining any nutrients or energy for development. The literature describes the mechanisms of microbial attack on DDT. According to most reports, DDT is reductively dechlorinated to DDD under reducing circumstances. Bacteria and fungi are known to metabolise DDT in this manner, and mechanisms for biodegradation via this method have been identified employing both. The metabolic pathways and mechanisms are detailed in the sections that follow. Unlike DDT, there have been few instances of microbial breakdown of DDE (Gohil and Ogram, 2020).

Reductive dechlorination

Under reducing circumstances, reductive dechlorination is the primary process for microbial conversion of both the o,p'-DDT and p,p'-DDT isomers of DDT to DDD. The process involves the replacement of aliphatic chlorine for a hydrogen atom.

Early investigations on the fate of DDT in rodents demonstrated that the indigenous microflora of animals were responsible for the conversion of DDT to DDD. Since then a number of microbes have been shown to be capable of converting DDT to DDD in pure culture. These include the bacteria *Escherichia coli, Enterobacter aerogenes, Enterobacter cloacae, Klebsiella pneumoniae, Pseudomonas aeruginosa, Pseudomonas putida. Bacillus sp.*, "Hydrogenomonas", and the fungi

Saccharomyces cerevisiae, Phanerochaete chrysosporium, and *Trichoderma viridae*. As the biotransformation of DDT to DDD by these microorganisms requires the presence of an alternative carbon (C) source, the process is co-metabolic.

DDT-metabolising microorganisms have been found in a variety of environments, including animal feces, soil, sewage, activated sludge, and marine and freshwater. There have been reports of aerobic biotransformation of DDT into DDD. Mechanisms for reductive dechlorination using transition metals and metal complexes as reductants. In most cases, the process entails a single electron transfer, the elimination of a chlorine ion, and the production of an alkyl radical. Transition metal complexes in microorganisms are linked to the active sites of electron transport molecules. *E. coli*'s reductive dechlorination of DDT, for example, needed flavin-adenine dinucleotide (FAD) and anaerobic conditions, whereas *E. aerogenes*' cytochrome oxidase was involved.

Under anaerobic conditions, DDD might undergo further metabolism. Pure culture studies with E. coli and E. aerogenes resulted in the isolation of DDD as a major metabolite and trace amounts of 1-chloro-2,2-bis (p-chlorophenyl) ethylene (DDMU), 1-chloro-2,2-bis (p-chlorophenyl) ethane (DDMS), 2,2-bis (p- chlorophenyl) ethylene (DDNU), 2,2-bis(p- chlorophenyl) ethanol (DDOH), bis (pchlorophenyl)-acetic acid (DDA), and 4,4' dichlorobenzophenone (DBP). Wedemeyer (1967) discovered that separate enzymes were involved in *E. aerogenes*' metabolism of DDT by using metabolic inhibitors as well as variations in pH and temperature. Degradation occurs through consecutive reductive dechlorination processes, resulting in DDNU, which is ultimately oxidized to DDOH. Further oxidation of DDOH produces DDA, which is decarboxylated to bis (p-chlorophenyl) methane (DDM). DDM is metabolized to DBP or, in the alternative, one of the aromatic rings is cleaved to create p-chlorophenylacetic acid (PCPA). Under anaerobic conditions, DBP was not further metabolized. SubbaRao and Alexander (1985) were able to validate Wedemeyer's (1967) pathway for the breakdown of DDT by the bacterium E. aerogenes by investigating the co-metabolism of DDT metabolites by a number of fungi. There is one report that a bacterium converts DDE to DDMU (Masse et al. 1989).

Alternating anaerobic and aerobic conditions promote reductive dechlorination and ring cleavage processes, respectively. The reaction mixture containing these metabolites was then subjected to aerobic conditions with a fresh culture of "Hydrogenomonas" sp. A metabolite known as PCPA was produced as a result of ring cleavage. Arthrobacter sp. converted this acid to p-chlorophenylglycoaldehyde. Since no PCPA was generated under anaerobic circumstances, these in vitro investigations indicate that ring cleavage processes required O2. DDT was not co-metabolized by "Hydrogenomonas" sp. in aerobic conditions.

Mineralisation of DDTr by ligninolytic fungi

Ligninolytic fungi have been demonstrated to exhibit biodegradative properties for a wide range of environmentally persistent chemicals, including DDT. This capability is linked to their ability to breakdown lignin. The majority of research on DDTr degradation by ligninolytic fungi has been conducted by Aust and colleagues, who used the white rot fungus *P. chrysosporium* and described the degradation of 14C-labelled DDT over 30 days in *P. chrysosporium* cultures lacking in nutritional nitrogen (N).

Approximately 50% of the DDT was transformed during this period with around 10% being mineralised and the remainder appeared as metabolites including dicofol, FW-152, and DBP. These results led to the proposal of a DDT degradation pathway involving oxidation to dicofol followed by dechlorination to FW-152 and subsequent breakdown via DBP. This pathway was thought to be controlled by the ligninase system of the fungus, as mineralisation and dicofol production were observed only after a lag phase during which ligninase production was established.

Naphthalene

Naphthalene has often been used as a model compound to investigate the ability of bacteria to degrade PAHs because it is the simplest and the most soluble PAH. Therefore, information of bacterial degradation of naphthalene has been used to understand and predict pathways in the degradation of three- or more ring PAHs (Mohapatra and Phale, 2021). Many bacteria that have been isolated and utilize naphthalene as a sole source of carbon and energy belong to the genera *Alcaligenes, Burkholderia, Mycobacterium, Polaromonas, Pseudomonas, Ralstonia, Rhodococcus, Sphingomonas,* and *Streptomyces.*

- » Degradation of naphthalene starts through the multicomponent enzyme, naphthalene dioxygenase, attack on the aromatic ring to form cis-(1R, 2S)-dihydroxy-1,2-dihydronaphthalene (cis-naphthalene dihydrodiol)
- » Naphthalene dioxygenase produces cis-naphthalene dihydrodiol, which is then dehydrogenated to 1,2-dihydroxynaphthalene by a cis-dihydrodiol dehydrogenase.
- » 1,2-dihydroxynaphthalene is then converted to salicylate using 2-hydroxy-2H-chromene-2-carboxylic acid, cis-o-hydroxybenzalpyruvate, and 2-hydroxybenzaldehyde. Additionally, 1,2-dihydroxynaphthalene is nonenzymatically reduced to 1,2-naphthaquinone.

» Salicylate is typically decarboxylated to catechol, which is further metabolized by ring fission in meta- and ortho-pathways
» Salicylate is converted to gentisate by salicylate-5-hydroxylase.

We isolated and studied the salicylate 1-hydroxylase from *Sphingomonas* sp. strain CHY-1. The bacterial breakdown of naphthalene has been thoroughly studied for the catabolic enzyme system expressed by the plasmid NAH7 in *Pseudomonas putida* G7. NAH7 has two operons that contain the structural genes required for naphthalene breakdown. One operon contains the gene for the upper catabolic pathway, which encodes the enzymes required for the conversion of naphthalene to salicylate. The second operon contains the gene for the lower catabolic pathway, which encodes the enzymes required for salicylate metabolism via the catechol meta-cleavage pathway to pyruvate and acetaldehyde. The whole sequencing structure of plasmid NAH7 (82,232 bp) has been established. In addition, the entire 83,042 bp sequence of the circular naphthalene degradation plasmid pDTG1 from Pseudomonas putida NCIB 9816-4 was determined. Aspartate 205 in the catalytic domain of naphthalene dioxygenase is an essential residue in the major electron transfer pathways to mononuclear iron in the active site.

8. 5. Plastics as a Xenobiotic in Environment

The word, plastic, was derived from the word ***'Plastikos'*** meaning ***'to mould'*** in Greek. Fossil fuels have compounds containing hydrogen and carbon (hydrocarbon) which act as building blocks for long polymer molecules. These building blocks are known as monomers, they link together to form long carbon chains called polymers.

Plastic is described as a material that comprises an organic compound with a high molecular weight. It is also defined as polymers with lengthy carbon chains. Carbon atoms form long-chain compounds.

Christian Schonbein, a famous German chemist, created plastic in 1846. Plastics were discovered accidently. Christian was experimenting in his kitchen when he accidentally spilled a mixture of nitric and sulphuric acid. He used a cloth to wipe the solution (a mixture of nitric and sulfuric acid) and then placed it over the stove. The cloth vanished after a while, and its name was derived from the plastic.

Plastic is a xenobiotic material and now it is accumulating with alarming rate in the environment. Plastic wastes continued to be resistant to natural degradation due to lack of proficient catabolic traits in the autochthonous microbial species. Xenobiotic compounds, on the other hand, refer to chemicals that are foreign to an organism's normal biochemistry and are not naturally occurring in the environment. Plastics can be considered xenobiotic compounds because they are human-made materials that do not occur naturally in ecosystems.

When plastics enter the environment, they can persist for long periods due to their chemical structure, which makes them resistant to degradation. As a result, they can accumulate in various ecosystems, including marine, freshwater, and terrestrial environments. Plastics can have detrimental effects on organisms and ecosystems, leading to various environmental issues. One of the major concerns associated with plastics as xenobiotic compounds is their potential for microplastic pollution. Microplastics are tiny plastic particles measuring less than 5 millimeters in size. They can originate from the breakdown of larger plastic items or be manufactured as microbeads for various consumer products. Microplastics can enter ecosystems through multiple pathways. They can be ingested directly by organisms, such as marine animals or birds, causing physical harm, intestinal blockage, or even death. Microplastics can also bioaccumulate in the food chain, potentially reaching higher trophic levels, including humans. The long-term impacts of microplastic ingestion on human health are still being studied, but there are concerns about potential health risks.

Furthermore, plastics can leach various additives and chemicals into the environment. These additives include plasticizers, flame retardants, stabilizers, and dyes. Some of these chemicals may have toxic properties and can pose risks to organisms exposed to them. For example, certain plasticizers, such as phthalates, have been associated with endocrine disruption and adverse effects on reproductive health in animals (Dutta and Choudhury, 2018).

Properties of plastics

Plastics are a group of synthetic materials that are versatile, durable, and widely used in various industries. Here are some key properties of plastics:

Lightweight: Plastics have a low density, making them lightweight materials.

Durable: Plastics are known for their durability and long lifespan. They can withstand harsh environmental conditions, including exposure to moisture, chemicals, and sunlight, without significant degradation.

Versatile: Plastics can be easily molded into different shapes and sizes, allowing for a wide range of applications. They can be molded, extruded, blow-molded, or formed through other processes, making them highly adaptable to various manufacturing techniques.

Resistance to Chemicals: Many plastics exhibit excellent resistance to chemicals, making them suitable for storing and transporting corrosive substances.

Electrical Insulation: Plastics are good electrical insulators, which mean they do not conduct electricity well.

Thermal Insulation: Plastics have relatively low thermal conductivity, providing insulation against heat transfer.

Transparency and Optical Clarity: Certain plastics, such as polycarbonate and acrylic, have excellent transparency and optical clarity, resembling glass.

Chemical Resistance: Plastics can have varying levels of resistance to different chemicals. Some plastics are highly resistant to specific chemicals, while others may be more susceptible to chemical attack.

Recyclability: Although not all plastics are easily recyclable, there are several types, such as PET (polyethylene terephthalate) and HDPE (high-density polyethylene), that are commonly recycled.

Type of plastics

Plastics can be broadly classified into several types based on their chemical composition and properties. Some common types of plastics are:

Polyethylene Terephthalate (PET): PET is commonly used for beverage bottles, food containers, and synthetic fibers. It is lightweight, transparent, and has good resistance to moisture and chemicals.

High-Density Polyethylene (HDPE): HDPE is known for its strength and durability. It is used in various applications, such as bottles for milk and detergent, pipes, and plastic bags. HDPE has good resistance to chemicals and impact.

Polyvinyl Chloride (PVC): PVC is a versatile plastic used in construction materials, pipes, electric cables, flooring, and medical devices. It can be rigid or flexible, depending on the additives used.

Low-Density Polyethylene (LDPE): LDPE is a flexible and transparent plastic commonly used for packaging films, plastic bags, and squeeze bottles. It has good resistance to moisture and chemicals.

Polypropylene (PP): PP is a versatile plastic with high resistance to heat, chemicals, and fatigue. It is used in a wide range of applications, including food containers, automotive parts, textiles, and packaging.

Polystyrene (PS): PS is a rigid and brittle plastic used for disposable food containers, packaging materials, insulation, and consumer electronics. It can be transparent or foam-like (Expanded Polystyrene, EPS).

Polyethylene (PE): PE is a generic term for different forms of polyethylene, including HDPE and LDPE. It is widely used in various applications, such as packaging, bottles, toys, and pipes.

Polyurethane (PU): PU is a versatile plastic used in foams, cushions, insulation, adhesives, and coatings. It can be flexible or rigid, depending on the formulation.

Polycarbonate (PC): PC is a strong and transparent plastic used in eyeglass lenses, safety goggles, CDs, and automotive components. It has excellent impact resistance.

Acrylonitrile Butadiene Styrene (ABS): ABS is a tough and rigid plastic commonly used in appliances, toys, computer parts, and automotive components.

Additionally, depending upon the physical properties of plastics, they are categorized into two categories i.e. Thermoplastic and thermosetting

i. *Thermoplastic*: Plastics that can be deformed easily upon heating and can be bent easily. Linear polymers and a combination of linear and cross-linked polymers come under thermoplastics. Example: PVC, nylon, polythene, etc.

ii. *Thermosetting*: Plastics that cannot be softened again by heating once they are moulded. Heavily cross-linked polymers come under the category of thermosetting plastics. Example: Bakelite, melamine, etc. Bakelite is used for making electrical switches whereas melamine is used for floor tiles.

8.5.1. Mechanism of biodegradation of plastics

The polymer is broken down into monomers, which are then mineralized. Because big polymers cannot pass through the cell membrane, they must first be depolymerized before entering the cell and being mineralized or biodegraded within it. This is observed in several synthetic polymers, such as polycaprolactone. Initially, different physical and biological factors work on the polymer to break it down. The physical forces include heating, cooling, freezing, thawing, wetting, drying and so on; they cause mechanical damage such as cracking of the polymer. The growth of fungi on the polymer can also cause swelling and bursting of the polymer to facilitate the fungi to penetrate into it.

Environmental degradation is initiated in synthetic polymers like poly carboxylates, polyethylene, poly terephthalate, poly lactic acids and their copolymers, poly dimethylsiloxanes, etc by abiotic hydrolysis.

Two types of enzymes are involved in biodegradation of polymers: *intracellular and extracellular depolymerases*. During degradation the exoenzymes or the extracellular depolymerases from the microorganisms convert the polymers into molecules having shorter chains. These molecules are monomers, dimers or oligomers that are small enough to pass through the semi permeable bacterial cell membrane. These are then utilised by the bacteria as sources of carbon and energy. This process is termed as depolymerisation. The process in which the end products are water, carbon dioxide or methane, is called mineralisation.

Availability of Oxygen

Aerobic microorganisms break down polymers in the presence of oxygen, producing microbial biomass, carbon dioxide, and water as byproducts. However, in anoxic or anaerobic environments, groups of anaerobic microbes break down the polymer, producing microbial biomass, water, methane, and carbon dioxide as byproducts (e.g. landfills, compost).

Factors affecting Biodegradation of Plastics

Depending on the characteristics of the soil, bacteria and fungus biodegrade plastics in different ways. The type of organism, the properties of the polymer, and the type of pre-treatment are the various factors that control biodegradation. The tacticity, mobility, crystallinity, molecular weight, kind of functional groups and other substituents in the structure, and additives or plasticizers added to the polymer are all considered features of the polymer. Smaller plastic particles were found to biodegrade more quickly than larger ones (Jadaun et al., 2022).

As a polymer's molecular weight increases, its biodegradability by microorganisms diminishes. Because the polymer must be absorbed into the bacterial cell membrane and degraded by cellular enzymes, an increase in molecular weight results in a decrease in polymer solubility, making it less favorable for microbial attack. Monomers, dimers, and oligomers are examples of repeating polymer units that are readily broken down and mineralized. Abiotic hydrolysis, photo-oxidation, and physical disintegration all promote biodegradation. These procedures increase the polymer's surface area and decrease its molecular weight, which promotes microbial breakdown.

Amorphous regions are more susceptible to microbial degradation than crystalline regains. The presence of a glucose source reduces the rate of biodegradation as glucose is a more preferred carbon source than plastic.

Microorganisms involved in Biodegradation of Plastics

Bacteria: Several bacterial species have been identified as plastic degraders. One well-known example is *Ideonella sakaiensis*, which was discovered in 2016 and is capable of breaking down the plastic PET (polyethylene terephthalate) commonly used in beverage bottles. Other bacteria like *Pseudomonas, Bacillus*, and *Rhodococcus* have also been found to degrade various types of plastics.

Fungi: Certain species of fungi have shown the ability to degrade plastics. For instance, the fungus *Aspergillus niger* has been found to degrade polyurethane (PU) foams. Other fungi like *Trichoderma*, *Penicillium*, and *Phanerochaete chrysosporium* have also demonstrated plastic-degrading capabilities.

Algae: Some species of algae have been reported to degrade plastics as well. For example, the green microalga *Chlorella vulgaris* has been found to break down low-density polyethylene (LDPE) films.

Enzymes: In addition to microorganisms, specific enzymes produced by these microorganisms play a crucial role in plastic degradation. For example, PETase (polyethylene terephthalate hydrolase) and MHETase (mono (2-hydroxyethyl) terephthalic acid hydrolase) are enzymes produced by *Ideonella sakaiensis* that target PET. These enzymes break down the molecular structure of the plastic, enabling its degradation.

8.6. Biodegradation of Polyolefins

Long carbon chains make up the backbone of synthetic polyolefins, which are inert. Microorganisms cannot break them down because of their high molecular weight and strong hydrophobicity. Nonetheless, following chemical and photodegradation, several microbes were able to degrade low molecular weight polyolefins. Polyolefins must undergo oxidation to increase their hydrophilic level and decrease their molecular weight and chain length in order to be sensitive to microbial breakdown.

Polyethylene

Starch and prooxidant can be added to polyethylene to make it more vulnerable to microbial breakdown. These are utilized in the production of biodegradable polyethylene. The starch makes polyethylene more hydrophilic, allowing amylase enzymes to catalyze it. This portion is easily broken down by microorganisms. Photodegradation and chemical degradation occur first when pro-oxidant is added to polyethylene.

Biotic and Abiotic hydrolysis

Plastics undergo biotic or abiotic degradation to produce monomeric or oligomeric compounds prior to biodegradation. While hydrophobic hydrocarbon polymers, like polyolefins, are resistant to hydrolysis and cannot be biodegraded, hydrobiodegradable polymers, including polyesters and thermoplastic starch, are broken down by biotic or abiotic hydrolysis. They can withstand oxidation as well. However, they can be made oxobiodegradable by using pro-oxidant chemicals.

Pro-oxidants that are based on metal combinations capable of yielding two metal ions having similar stability and having oxidation number differing by only one unit are the most active or effective. E.g. Mn^{2+}/Mn^{3+}. The material uses Oxygen from the atmosphere to degrade by a free radical chain reaction. Hydroperoxidases are formed as the primary product; they either thermolyse or photolyse under the catalytic action of a pro-oxidant. This leads to chain scission that produces low molecular mass oxidation products such as carboxylic acids, alcohols, ketones and hydrocarbon waxes.

Microorganisms benefit from hydrophilic surface change brought on by peroxidation. The low molecular mass oxidation products can be bioassimilated by them. Prooxidants thereby facilitate polyethylene's thermo-oxidative breakdown. The rate of degradation is influenced by temperature, the kind and quantity of additives, the partial pressure of oxygen, and the kind of polymer. The most significant factor influencing polyethylene deterioration is temperature. The rate of degradation can be significantly slowed down by low partial pressure of oxygen.

Polyvinylchloride (PVC)

PVC is durable and absorbs little moisture. It resists abrasion. Chemical and photodegradation are two ways that PVC can be broken down, but there aren't many reports on PVC biodegradation. It has various commercial and industrial applications such rigid pipes, shoe bottoms, garden hoses, textiles, electrical wire insulation, pipes and fittings, synthetic leather products and floor coverings.

Polystyrene

Styrene, benzene, toluene, and acrolein are results of the heat or chemical decomposition of polystyrene, a synthetic material. Although biodegradation of polystyrene's monomer styrene has been documented a few times, there are currently very few reports of biodegradation of polystyrene. Because of its qualities like lightweight, rigidity, and thermal insulation, it has numerous commercial and industrial uses, including the production of disposable cups, packaging materials, electronics, and lab equipment.

8.7. Biodegradation of Natural Plastics (Polyhydroxylkanoates)

Under nutrient-limited situations, bacteria make and store PHA in their cells. When the nutrients are available, this PHA is broken down and metabolized. However, the degradation of PHA in the environment is not guaranteed by its storage capacity. Polymers must be broken down into their smaller subunits or monomers in order to cross the semi-permeable cellular membrane. In order to accomplish this, bacteria have developed extracellular hydrolases that can break down polymers into their corresponding hydroxyl acid monomers. R-3-hydroxybutyric acid is produced when PHB is hydrolyzed, while 3-hydroxybutyrate and 3-hydroxyvalerate are produced when PHBV is hydrolyzed. These monomers are tiny enough to fit through the cell wall and are soluble in water. After entering the cell, they are broken down by the TCA cycle and beta oxidation to produce carbon dioxide and water under aerobic circumstances. However, methane is created along with the other products when there is no oxygen present or when the environment is anaerobic. With the exception of P. lemoignei, which produces at least seven distinct extracellular PHA depolymerases with somewhat diverse biochemical characteristics, the majority of PHA-degrading bacteria only produce one PHA depolymerase. This procedure typically produces no hazardous byproducts. Higher creatures also include 3-hydroxybutyrate in their blood plasma. PHA has several medical uses, including surgical pins, blood vessel and bone replacement, sutures, and long-term controlled drug release, because it is completely safe.

Biodegradation of Polyurethane

Polyurethanes belong to a synthetic class of polymers which are made of urethane linkages. An alcohol functional group and an isocyanate functional group react together to form a urethane linkage. Polyurethanes have a wide variety of commercial applications. They are used in paints, thermal insulating materials. Polyurethane is susceptible to biodegradation despite its xenobiotic origins. A bacterial strain, *Comamon asacidovorans* TB35, utilises polyester polyurethane polymers containing

Polydiethylene glycol Adipate as a carbon source. It has polyester polyurethane (PUR)-degrading enzyme, PUR esterase, bound to its cell surface that can degrade Polyurethane. Biodegradation by other bacteria is incomplete as they are not able to use this as the sole carbon source; they require additional carbon sources as well. Biodegradation of polyurethane is indicated by an increase in the number and size of pores on its surface that can be seen under an electron microscope.

Table. 8.1. Microorganism involved in bioremediation of plastics

Plastic	Microorganism	Reference
Polyethylene	*Penicillium simplicissimum YK*	Yamada-Onodera et al., 2001
	Rhodococcus rubber	Sivan et al. (2006); Gilan et al., 2004
	Brevibacillus borstelensis	Hadad et al. (2005)
Polyurethane	*Curvularia senegalensis*	Howard (2002)
	Fusarium solani	
	Aureobasidium pullulans	
	Cladosporium sp.	
	Pseudomonas chlororaphis	Zheng et al. (2005)
	Comamonas acidovorans TB-35	Akutsu et al., 1998
Polyvinylchloride	*Pseudomonas putida AJ*	Anthony et al. (2004)
	Ochrobactrum TD	
	Pseudomonas fluorescens B-22	Mogil`nitskii et al. (1987)
	Aspergillus niger	
	van Tieghem F-1119	
Plasticized Polyvinyl chloride	*Aureobasidium pullulans*	Webb et al. (2000)
BTA-copolyester	*Thermomonspora fusca*	Kleeberg et al. (1998)

8.6. Microplastics

Microplastics are tiny plastic particles that result from both commercial product development and the breakdown of larger plastics. Officially, they are defined as plastics less than five millimeters (0.2 inches or 500μm) in diameter. Microplastics, were first described in the early 1970s and are widespread in the ocean. In 2004, Professor Richard Thompson and his team were the first to show that microplastic particles have accumulated in oceans. Microplastics can indeed be considered xenobiotics. Microplastics have been detected in marine organisms from plankton to whales, in commercial seafood, and even in drinking water. Alarmingly, standard water treatment facilities cannot remove all traces of microplastics. To further complicate matters, microplastics in the ocean can bind with other harmful chemicals before being ingested by marine organisms (Wu et al., 2017).

Microplastics can enter the environment through various sources, including plastic waste, microbeads in personal care products, and the breakdown of synthetic fibers. Due to their small size and persistent nature, microplastics can have detrimental effects on organisms and ecosystems. They are often ingested by a wide range of organisms, from microscopic plankton to large marine mammals. The presence of microplastics in the gastrointestinal tracts of organisms can lead to physical blockages, reduced nutrient uptake, and damage to internal organs. Furthermore, microplastics can act as vectors for other harmful substances. They have a high surface area to volume ratio and can absorb and concentrate other pollutants present in the environment, such as heavy metals, pesticides, and hydrophobic organic compounds. This can result in the transfer of these toxic substances to organisms when they ingest microplastics.

The impact of microplastics as xenobiotics extends beyond individual organisms. They can disrupt ecosystem functioning, alter food webs, and affect the overall health of aquatic and terrestrial environments. The long-term ecological consequences of microplastic pollution are still being studied, but evidence suggests that they pose significant risks to biodiversity and ecosystem stability.

There are two categories of microplastics: primary and secondary.Primary microplastics are tiny particles designed for commercial use, such as cosmetics, as well as microfibers shed from clothing and other textiles, such as fishing nets. Secondary microplastics are particles that result from the breakdown of larger plastic items, such as water bottles. This breakdown is caused by exposure to environmental factors, mainly the sun's radiation and ocean waves.

Primary and secondary microplastics can be further categorized:

Primary Microplastics:

i. *Microbeads*: These are tiny plastic particles intentionally manufactured and added to personal care and cosmetic products, such as exfoliating scrubs, toothpaste, and body wash. They are designed to provide abrasion or texture but are often too small to be effectively filtered out by wastewater treatment plants. As a result, they can enter water bodies directly.

ii. *Nurdles*: Also known as "pre-production pellets" or "plastic resin pellets," nurdles are small plastic pellets used as raw materials in the manufacturing of plastic products. They can escape during transportation, handling, and manufacturing processes and end up in the environment.

iii. *Microfibers:* These are microscopic fibers that shed from synthetic textiles like polyester, nylon, and acrylic. When we wash clothes made of these materials, microfibers are released into wastewater, and they can eventually find their way into aquatic environments.

iv. Microplastic granules and fragments: These are small fragments or granular forms of plastic resulting from the degradation and fragmentation of larger plastic items, such as bottles, bags, and packaging materials. These fragments can be further broken down by physical, chemical, and biological processes into smaller microplastics.

Secondary Microplastics:

i. *Weathering and fragmentation*: Over time, larger plastic items exposed to sunlight, wind, waves, and other environmental stressors undergo weathering and mechanical degradation, leading to the formation of secondary microplastics.

ii. *Degradation of microbeads and microfibers*: Microbeads and microfibers can also break down into smaller particles due to environmental processes, further contributing to secondary microplastic pollution.

iii. *Plastic debris breakdown*: Plastic waste that accumulates in the environment, such as plastic bags, bottles, and fishing gear, can degrade into smaller pieces through natural weathering processes, forming secondary microplastics.

Impact of microplastics

Microplastics have a range of impacts on the environment, ecosystems, and organisms. Here are some key impacts associated with microplastic pollution:

i. *Ecosystem Disruption*: Microplastics can disrupt ecosystem dynamics and functioning. They can accumulate in various habitats, including freshwater systems, oceans, and terrestrial environments. Their presence can affect nutrient cycles, alter species composition, and disrupt food webs. This can have cascading effects on ecosystem health and stability.

ii. *Ingestion and Physical Effects*: Microplastics are often ingested by a wide range of organisms, from tiny zooplankton to larger marine mammals. The ingestion of microplastics can lead to physical damage in the digestive system, blockage of digestive tracts, reduced nutrient absorption, and internal injuries. These effects can impair growth, reproduction, and overall health of affected organisms.

iii. *Toxicological Effects*: Microplastics can absorb and concentrate toxic pollutants present in the environment, such as heavy metals and organic compounds. When organisms ingest microplastics laden with these pollutants, they can be exposed to higher concentrations than they would encounter naturally. This can lead to toxicological effects, including disrupted hormone regulation, impaired immune function, and increased vulnerability to diseases.

iv. *Distribution and Bioaccumulation*: Microplastics can be transported over long distances by wind, currents, and wildlife, leading to their widespread distribution. They can accumulate in sediments, on the seafloor, and in various organisms throughout the food chain. Microplastics have been found in organisms ranging from filter-feeding invertebrates to top predators, resulting in bioaccumulation and biomagnification of these particles and associated pollutants.

v. *Behavioral and Reproductive Effects*: Microplastics can influence the behavior and reproductive processes of organisms. For example, some studies suggest that exposure to microplastics can alter feeding behaviors, reduce reproductive success, and impact the development of offspring in various marine species. These effects can have long-term consequences for population dynamics and ecosystem health.

vi. *Economic Implications*: Microplastic pollution also has economic implications. It can impact fisheries and aquaculture by affecting the abundance and health of commercially important species. It can also impact industries such as tourism, as the presence of microplastics on beaches and in recreational areas can diminish their aesthetic value and appeal.

Several potential strategies have shown promise. Here are a few examples:

i. *Microbial degradation*: Some microorganisms, such as bacteria and fungi, have the ability to degrade certain types of plastics. Researchers have identified microorganisms capable of utilizing plastics as a carbon source and breaking them down through enzymatic activities. This microbial degradation could potentially be harnessed to develop bioremediation strategies for microplastics.

ii. *Biofilm formation*: Biofilms are communities of microorganisms that adhere to surfaces. They have the potential to colonize microplastic particles and facilitate their degradation. Biofilm formation on microplastics can enhance microbial degradation processes and promote the breakdown of plastics into smaller fragments, which may be more susceptible to further degradation.

iii. *Enzymatic treatment*: Enzymes produced by microorganisms or engineered enzymes with plastic-degrading properties are being explored for their potential in breaking down microplastics. Enzymatic treatment has shown promise in laboratory settings for degrading certain types of plastics, although further research is needed to optimize the process for real-world applications.

iv. *Marine organisms*: Some marine organisms, such as certain species of bacteria, algae, and marine worms, have been found to interact with and potentially degrade microplastics. Researchers are studying the mechanisms employed by these organisms to better understand their potential for bioremediation purposes.

Metals as xenobiotic compound in environment and their bioremediation

Metals are naturally occurring elements that forms positive ions and has metallic bonds. Common example of metal includes iron, copper, silver, mercury, lead, aluminum, gold, platinum, zinc, nickel and tin. When they are present in the low amount in environment they do not cause any adverse because they are required to all the living organism including microorganism as a supplement in their food which act as a cofactors of enzymes to mediate the metabolic process, indispensable for cellular energy production and growth. However, when they are released in the excess amount in the environment by anthropogenic activities they are not readily metabolized or eliminated by living organisms, they impose adverse effect to organism at cellular and morphological level.

But, some metals with relatively high densities, atomic weights, or atomic numbers are known as heavy metals. A density of heavy metal is more than 5 g/cm3. Examples of heavy metals are arsenic (As), cadmium (Cd), chromium (Cr), copper (Cu), lead (Pb), manganese (Mn), mercury (Hg), nickel (Ni), uranium (U) etc. They can be released into the environment through industrial activities, mining, improper waste disposal, and other human activities. These metals can accumulate in ecosystems and cause various harmful effects on organisms, including developmental abnormalities, organ damage, and even death.

Some metals which are having properties of both metals and non-metals are known as metalloids. The six commonly recognized metalloids are boron, silicon, germanium, arsenic, antimony and tellurium. Five elements are less frequently so classified: carbon, aluminium, selenium, polonium and astatine. On a standard periodic table, all eleven elements are in a diagonal region of the p-block extending from boron at the upper left to astatine at lower right. Some periodic tables include a dividing line between metals and nonmetals, and the metalloids may be found close to this line. One of the most well-known metalloids is arsenic (As), which can be found in both organic and inorganic forms. Inorganic arsenic compounds are particularly toxic and can contaminate soil, water, and air through natural processes or human activities such as mining and the use of certain pesticides. Arsenic exposure can lead to various health issues, including skin lesions, cancers, and neurological disorders.

Nanoparticles of certain metals, such as silver (Ag), copper (Cu), and zinc (Zn), are used in various industrial and consumer products due to their antimicrobial and other unique properties. However, when these nanoparticles are released into the environment, they can pose risks to ecosystems and organisms. Their small size and large surface area make them highly reactive and potentially toxic to living organisms, affecting cellular processes and even leading to ecological disruptions. Rare earth elements (REEs), which include elements such as cerium (Ce), neodymium (Nd), and lanthanum (La), are used in various technologies, including electronics and renewable energy systems. Improper disposal of electronic waste or mining activities can lead to the release of REEs into the environment. Although the toxicity of REEs varies, some of these elements can accumulate in living organisms and have adverse effects on their growth, reproduction, and overall health.

8.7. Hydrocarbons as Xenobiotics

Hydrocarbons are organic compounds consisting entirely of hydrogen and carbon atoms and are commonly found in fossil fuels, petroleum, and various industrial chemicals. Hydrocarbons can enter the environment through natural processes such as the seepage of crude oil from underground reservoirs or through human activities like oil spills, industrial discharges, and improper disposal of petroleum products. When released into ecosystems, hydrocarbons can have significant impacts on organisms and the environment. Hydrocarbons, especially certain types such as polycyclic aromatic hydrocarbons (PAHs) and volatile organic compounds (VOCs), can be toxic to organisms. They can disrupt physiological processes, impair reproductive and developmental functions, and cause acute or chronic health issues. The toxicity of hydrocarbons depends on their chemical composition, concentration, and exposure duration. Hydrocarbons, particularly persistent ones, have the potential to bioaccumulate in organisms. This means that they can accumulate in the tissues of organisms over time, as they are absorbed faster than they are metabolized or excreted. In food chains, hydrocarbons can undergo biomagnification, where their concentrations increase as they move up the trophic levels, leading to higher exposure levels in top predators.

Some hydrocarbons, such as certain PAHs, can persist in the environment for extended periods. They may resist degradation by natural processes and can persist in soil, sediments, and water bodies. This persistence increases the likelihood of long-term exposure to organisms and ecosystems. Hydrocarbon pollution can have severe ecological consequences. It can lead to the loss of biodiversity, habitat degradation, and disruption of ecological interactions. Aquatic ecosystems, in particular, are highly vulnerable to hydrocarbon contamination, as it can affect water quality, aquatic organisms, and the overall health of the ecosystem. Large-scale oil spills, such as those resulting from tanker accidents or offshore drilling mishaps, can have devastating effects on marine environments. The oil spreads over the water surface, forming slicks that can contaminate coastlines, coral reefs, and other sensitive habitats. Oil spills can lead to the death of marine organisms, damage to their habitats, and long-term ecological repercussions.

Types of Hydrocarbons

- *Saturated Hydrocarbons*: In these compounds, carbon-carbon atoms and carbon-hydrogen atoms are held together by single bonds. These single-bonded compounds are the simplest hydrocarbons. These types of hydrocarbons don't have double or triple bonds. In terms of hybridization, they have Sp^3 hybridised carbon atoms with no Sp^2 or Sp hybridised carbon atoms. They are together called alkanes which have a general formula of C_nH_{2n+2}. For example, Methane (CH_4), ethane (C_2H_6), propane (C_3H_8), and butane (C_4H_{10}).
- *Unsaturated Hydrocarbons*: These compounds consist of a single, double or triple bond between carbon-carbon atoms. The double-bonded compounds are called alkenes, and the triple-bonded compounds are called alkynes. The general formula for alkenes is $C_nH_{2n,}$ and for alkynes, the general formula is C_nH_{2n-2}. For example, For Ethene (C_2H_4), benzene (C_2H_6), acetic acid ($C_2H_4O_2$).
- *Cycloalkanes*: These hydrocarbons possess one or multiple carbon rings. The hydrogen atom is attached to the carbon ring. For example, cyclopentane, Cyclobutane, cyclohexane, and cycloheptane, cyclooctane, etc.
- *Aromatic Hydrocarbons*: They are also called arenes. Arenes are compounds

which consist of at least one aromatic ring. For example, Methylbenzene, Naphthalene, Phenanthrene, Trinitrotoluene, and o-dihydroxybenzene.

Toxicity of hydrocarbon in environment

They are widely used in various industrial processes and products, and they can also be found naturally in crude oil and natural gas. When hydrocarbons are released into the environment, either through accidental spills or intentional discharge, they can have toxic effects as xenobiotics.

The toxicity of hydrocarbons depends on various factors, including their chemical structure, concentration, duration of exposure, and the specific organism or ecosystem being affected. Here are some ways in which hydrocarbons can be toxic to the environment:

» Acute Toxicity: High concentrations of hydrocarbons, particularly volatile ones such as gasoline or certain solvents, can cause acute toxicity when organisms are exposed to them. Inhalation or direct contact with hydrocarbon vapors or liquids can lead to respiratory distress, skin irritation, eye damage, and in severe cases, systemic effects and organ damage.

» Carcinogenicity: Some hydrocarbons, such as polycyclic aromatic hydrocarbons (PAHs), are known or suspected to be carcinogenic. Chronic exposure to these compounds can increase the risk of developing cancer in both humans and animals.

» Bioaccumulation: Hydrocarbons can accumulate in the tissues of organisms through the process of bioaccumulation. This is especially true for hydrophobic compounds, such as certain chlorinated hydrocarbons and some long-chain hydrocarbons. As they move up the food chain, these hydrocarbons can become more concentrated, posing a greater risk to predators or higher-level consumers.

» Disruption of Ecosystems: Hydrocarbons released into aquatic environments can have severe impacts on marine life. They can interfere with the respiratory systems of fish and other aquatic organisms, leading to suffocation or reduced oxygen uptake. Hydrocarbons can also contaminate habitats and disrupt the balance of ecosystems, affecting the growth and reproduction of plants and animals.

» Environmental Persistence: Some hydrocarbons, especially those with complex chemical structures or resistance to degradation, can persist in the environment for long periods. This persistence can lead to chronic exposure and long-term effects on ecosystems.

Toxicity of hydrocarbons in aquatic ecosystem

Hydrocarbons when released into aquatic environments, hydrocarbons can have significant adverse impacts on aquatic life. There are two main types of hydrocarbons relevant to aquatic toxicity:

Petroleum Hydrocarbons:

These are hydrocarbons derived from crude oil and its refined products. They are commonly released into aquatic ecosystems through oil spills, leakage from ships, offshore drilling, and other industrial activities. Petroleum hydrocarbons can be broadly categorized into two groups:

(i) Non-Polar Hydrocarbons: These hydrocarbons, such as alkanes, are relatively less toxic but can still harm aquatic life by coating the surfaces of organisms and interfering with oxygen exchange, feeding, and other vital functions. They also have the potential to accumulate in the food chain, leading to biomagnification.

(ii) Aromatic Hydrocarbons: These include compounds like benzene, toluene, ethylbenzene, and xylene (BTEX). Aromatic hydrocarbons are more toxic than non-polar hydrocarbons and can cause severe effects on aquatic organisms, including damage to their tissues, impaired reproduction, and developmental abnormalities.

Polycyclic Aromatic Hydrocarbons (PAHs): PAHs are a class of hydrocarbons formed during incomplete combustion of organic matter. They are widespread pollutants in aquatic environments and are commonly associated with industrial activities, vehicle emissions, and urban runoff. PAHs are highly toxic to aquatic organisms due to their persistence and ability to be absorbed by organisms. They can cause various health issues, such as cancer, in fish, invertebrates, and other aquatic life forms.

Examples of hydrocarbon toxicity in aquatic ecosystems:

i. *Deepwater Horizon Oil Spill*: One of the most significant examples of hydrocarbon toxicity in aquatic ecosystems is the Deepwater Horizon oil spill that occurred in the Gulf of Mexico in 2010. The spill released millions of barrels of crude oil, leading to extensive damage to marine life, including fish, birds, and marine mammals. The toxic effects of petroleum hydrocarbons impacted the reproductive success and survival of many species, and it took years for the ecosystem to recover fully.

ii. *Exxon Valdez Oil Spill*: Another infamous example is the Exxon Valdez oil spill that occurred in 1989 in Prince William Sound, Alaska. The spill resulted in the release of over 11 million gallons of crude oil, affecting vast stretches of coastline and marine habitats. It had severe consequences for various aquatic organisms, such as seabirds, sea otters, and marine invertebrates, causing both immediate and long-term harm to the ecosystem.

iii. *Runoff from Urban Areas*: Urban runoff, which includes pollutants from roads, parking lots, and industrial areas, is a significant source of hydrocarbon contamination in aquatic ecosystems. These pollutants, including PAHs, can negatively impact aquatic organisms living in nearby rivers, streams, and lakes, leading to reduced biodiversity and ecosystem health.

Hydrocarbon toxicity in terrestrial ecosystem

Hydrocarbon toxicity in terrestrial ecosystems refers to the harmful effects of hydrocarbons on the various plants, animals, and microorganisms living on land. Like in aquatic ecosystems, hydrocarbons can originate from both natural sources (e.g., forest fires, volcanic activity) and human activities (e.g., industrial processes, vehicle emissions, oil spills, and leaks). When hydrocarbons are released into the environment, they can have significant adverse impacts on terrestrial life, vegetation, and soil health. The extent of hydrocarbon toxicity in terrestrial ecosystems depends on factors such as the type and concentration of hydrocarbons, the duration of exposure, the sensitivity of the affected organisms, and the overall resilience of the ecosystem. The following are some examples of hydrocarbon toxicity in terrestrial ecosystems:

i. *Soil Contamination*: Hydrocarbons, particularly petroleum-based products like gasoline, diesel, and crude oil, can contaminate the soil when spilled or leaked from storage tanks, pipelines, or industrial facilities. These hydrocarbons can persist in the soil for extended periods, affecting soil fertility, microbial activity, and the health of plants. The presence of hydrocarbons in the soil can inhibit seed germination, root growth, and nutrient uptake by plants, leading to reduced plant productivity and biodiversity.

ii. *Air Pollution*: Hydrocarbons released into the atmosphere from various sources can contribute to air pollution. In urban areas, vehicle emissions are a significant source of hydrocarbon pollution. When hydrocarbons combine with other pollutants like nitrogen oxides and sunlight, they can form ground-level ozone and smog. These pollutants can have harmful effects on plants, causing leaf damage, reduced photosynthesis, and increased susceptibility to diseases and pests.

Example: Cities with heavy traffic and industrial activity, such as Los Angeles, experience high levels of urban air pollution. The combination of vehicle emissions and industrial hydrocarbon releases contributes to the formation of smog, which negatively impacts plant health in the surrounding areas.

iii. *Forest Fires*: Forest fires are a natural source of hydrocarbon release in terrestrial ecosystems. When vegetation burns, it releases volatile organic compounds (VOCs), which include various hydrocarbons. While some plants have adapted to fire and even require it for seed germination, intense and uncontrolled fires can lead to significant ecological disruption, loss of habitat, and even the release of harmful air pollutants.

Example: The wildfires that occur in various regions worldwide, such as the wildfires in California, Australia, and the Amazon rainforest, release substantial amounts of hydrocarbons and other pollutants into the atmosphere, affecting both terrestrial and aquatic ecosystems.

iv. Oil Spills on Land: Oil spills that occur on land, such as in industrial sites or during transportation, can result in the direct contamination of terrestrial ecosystems. The spilled oil can coat vegetation, soil, and other surfaces, leading to toxic effects on plants and wildlife. These spills can be especially harmful when they occur in sensitive habitats like wetlands and wildlife reserves.

Example: In 2016, a pipeline rupture in the U.S. state of North Dakota led to the release of more than 176,000 gallons of crude oil into a creek, causing significant damage to the surrounding terrestrial ecosystem. The spill affected farmland, wildlife habitats, and water sources, posing risks to the local flora and fauna.

v. Agricultural Runoff: Pesticides and herbicides used in agricultural practices can contain hydrocarbons and other chemicals. Improper use or runoff from agricultural fields can lead to the contamination of nearby terrestrial ecosystems, impacting non-target plants, insects, and soil organisms.

Example: The use of herbicides like glyphosate in agriculture has been associated with environmental concerns, including soil and water contamination, which can indirectly impact terrestrial ecosystems.

8.8. Biodegradation of Hydrocarbon

Biodegradation of hydrocarbons by natural populations of microorganisms represents one of the primary mechanisms by which petroleum and other hydrocarbon pollutants are eliminated from the environment. It plays a significant role in the natural attenuation of hydrocarbon contaminants in the environment. Here are some key points regarding the biodegradation of hydrocarbons:

Hydrocarbon-Degrading Microorganisms

Certain bacteria, fungi, and other microorganisms have the metabolic capability to utilize hydrocarbons as a source of carbon and energy. These hydrocarbon-degrading microorganisms possess specific enzymes that enable them to break down the complex hydrocarbon molecules into smaller, more manageable compounds. Based on the number of published reports, the most important hydrocarbon-degrading bacteria in both marine and soil environments are *Achromobacter, Acinetobacter, Alcaligenes, Arthrobacter, Bacillus, Flavobacterium, Nocardia*, and *Pseudomonas spp.* and the *coryneforms*; the importance of hydrocarbon-degrading strains of the salt requiring *Vibrio spp.* is limited to the marine 'environment. Among the fungi, *Aureobasidium, Candida, Rhodotorula*, and *Sporobolomyces spp.* are the most common marine isolates and *Trichoderma* and *Mortierella* spp. are the most common soil isolates. Hydrocarbon-degrading *Aspergillus* and *Penicillium* spp. have been frequently isolated from both environments.

Aerobic Biodegradation mechanism of hydrocarbon:

Aerobic biodegradation is the process by which microorganisms break down hydrocarbons in the presence of oxygen. This natural process occurs in various environments, such as soil, water, and sediments, where oxygen is available. The breakdown of hydrocarbons in an aerobic environment follows a series of steps, and it involves the participation of different types of microorganisms.

The mechanism of aerobic biodegradation of hydrocarbons can be summarized in the following steps:

Introduction and Attachment: Aerobic hydrocarbon-degrading microorganisms, such as bacteria and certain fungi, colonize the hydrocarbon-contaminated area. These microorganisms possess specific enzymes capable of breaking down hydrocarbon molecules.

Enzymatic Breakdown: The microorganisms secrete enzymes known as oxygenases, which are responsible for initiating the breakdown of hydrocarbons. The oxygenases introduce molecular oxygen (O_2) into the hydrocarbon molecules, leading to the formation of reactive oxygen species (ROS). These ROS play a key role in the oxidation of the hydrocarbons, making them more water-soluble and easier to degrade.

Initial Oxidation: During the initial oxidation step, the microorganisms target the alkane and alkene components of the hydrocarbons. Alkanes, which are saturated hydrocarbons, are sequentially oxidized at the terminal carbon atom to form alcohols, aldehydes, and finally, carboxylic acids. Alkenes, which are unsaturated hydrocarbons with at least one double bond, are also oxidized to form epoxides.

Cleavage of Molecules: Further enzymatic reactions lead to the cleavage of hydrocarbon molecules into smaller fragments. These smaller fragments are more amenable to microbial degradation and assimilation.

Central Metabolic Pathways: The smaller hydrocarbon fragments enter the central metabolic pathways of the microorganisms. In bacteria, common pathways like the β-oxidation pathway and the Krebs cycle (citric acid cycle) are involved in the breakdown of hydrocarbons. These pathways yield energy and building blocks for the growth and reproduction of the microorganisms.

Complete Biodegradation: Through successive oxidation reactions and metabolic processes, the microorganisms ultimately convert the hydrocarbons into carbon dioxide (CO_2) and water (H_2O), along with some biomass. This process is known as complete biodegradation, and it results in the complete mineralization of the hydrocarbons.

Table 8.2.: Bacteria and fungi with hydrocarbons-biodegradation potential

Microbial genera	Hydrocarbon substrate
Pseudomonas	benzene, crude oil, anthracene, PCBs, diesel
Acinetobacter	crude oil
Bacillus	alkanes, crude oil, phenol, cresol
Kocuria	TPH, diesel
Alcaligenes	alkylbenzene sulfonates, PAHs, PCBs
Xanthomonas	PAHs
Rhodococcus	naphthalene, biphenyl
Arthrobacter	benzene, PAHs
Serratia	diesel

Corynebacterium	phenoxyacetates
Candida	PCBs, aldehydes, PAHs
Graphium	crude oil
Trichoderma	crude oil
Penicillium	crude oil
Phaenerochaete	halogenated hydrocarbons

i. Aerobic Biodegradation mechanism of hydrocarbon:

Anaerobic biodegradation is the process by which microorganisms break down hydrocarbons in the absence of oxygen. This natural process occurs in environments with limited or no oxygen, such as waterlogged soils, deep sediments, and the subsurface of certain environments. Unlike aerobic biodegradation, which relies on oxygen as an electron acceptor, anaerobic biodegradation involves the use of alternative electron acceptors during microbial metabolism. The mechanism of anaerobic biodegradation of hydrocarbons can be summarized in the following steps:

Hydrocarbon Hydrolysis: Anaerobic hydrocarbon-degrading microorganisms, such as certain bacteria and archaea, first need to initiate the process by breaking down large hydrocarbon molecules into smaller fragments. This initial step is called hydrolysis and involves the addition of water to the hydrocarbon bonds, resulting in the cleavage of the hydrocarbon molecules.

Fermentation: After hydrolysis, the smaller hydrocarbon fragments enter the fermentation process. During fermentation, the microorganisms partially oxidize the hydrocarbon fragments, producing simpler organic compounds, such as short-chain fatty acids, alcohols, and hydrogen gas (H_2).

Anaerobic Respiration: The key characteristic of anaerobic biodegradation is the use of alternative electron acceptors other than oxygen during respiration. In the absence of oxygen, microorganisms utilize various electron acceptors, depending on their availability in the environment. Common electron acceptors in anaerobic environments include nitrate (NO^{3-}), sulfate (SO_4^{2-}), carbon dioxide (CO_2), and even certain metal ions like iron (Fe^{3+}) and manganese (Mn^{4+}).

Methanogenesis: One essential and prevalent pathway in anaerobic biodegradation is methanogenesis. Certain archaea, known as methanogens, utilize the hydrogen gas (H_2) produced during fermentation and reduce carbon dioxide (CO_2) or other simple organic compounds to produce methane (CH_4). Methanogenesis is a crucial step in anaerobic degradation as it allows the transformation of the remaining organic compounds into a less toxic and more stable end product, methane gas.

Complete Biodegradation: As the process progresses, the hydrocarbon fragments are further broken down through various anaerobic metabolic pathways, such as syntrophic relationships between different microorganisms. Eventually, the hydrocarbons are converted into methane (CH_4) and carbon dioxide (CO_2) or other simpler end products.

However, syntrophy—a process in which the breakdown of a substrate by one microbe depends on the activity of another bacterium that maintains intermediate products like formate and H_2 at low concentrations is more frequently used in the anaerobic degradation of hydrocarbons. Reactions that would otherwise be thermodynamically unfavorable are driven by low product concentrations. Because using oxygen as a terminal electron acceptor is more energetically advantageous, syntrophy is more prevalent in anaerobic environments.

Since methanogens (archaea that make methane) can only metabolize simple substrates like acetate and hydrogen, syntrophic activities are essential necessary for full degradation to methane and carbon dioxide. Depending on the conditions and substrates available, a particular ecosystem may contain several syntrophic interactions. Primary degraders like Peptococcaceae and Clostridium break down hydrocarbons into intermediates like H2 and acetate in methanogenic settings, where all other electron acceptors have been depleted. Methanogens then consume these intermediates. In order to activate hydrocarbons without the need for molecular oxygen (O2), anaerobic microorganisms employ a variety of techniques. Each tactic, which works with both aliphatic and aromatic molecules, is covered in detail below.

Anaerobic biodegradation is generally slower than aerobic biodegradation due to the lower energy yields from the use of alternative electron acceptors. However, anaerobic processes play a crucial role in the natural degradation of hydrocarbons, especially in environments where oxygen is limited or absent, such as in deep sediments and the subsurface. Understanding the mechanisms of anaerobic biodegradation is essential for managing and remediating hydrocarbon-contaminated sites in a variety of environmental contexts.

Factors influencing bioremediation and biodegradation of hydrocarbon

Bioremediation and biodegradation of hydrocarbons are complex processes influenced by various factors. Understanding these factors is crucial for effectively utilizing microbial activity to clean up hydrocarbon-contaminated environments. Some of the key factors that affect bioremediation and biodegradation of hydrocarbons are as follows:

1. *Oxygen Availability*: Oxygen is a critical factor in determining whether biodegradation occurs aerobically or anaerobically. Aerobic biodegradation, which requires oxygen, is generally faster and more efficient than anaerobic biodegradation. Oxygen availability affects the types of microbial communities present, as different microorganisms thrive under aerobic or anaerobic conditions.

2. *Hydrocarbon Type and Complexity*: The structure and complexity of hydrocarbons influence their biodegradability. Simple hydrocarbons, such as straight-chain alkanes, are generally more easily biodegraded than complex aromatic hydrocarbons like polycyclic aromatic hydrocarbons (PAHs). Microorganisms have specific enzymes that target different types of hydrocarbons, and the presence of certain functional groups or branching can hinder biodegradation.

3. *Microbial Population*: The presence and abundance of hydrocarbon-degrading microorganisms in the contaminated environment significantly affect the rate and efficiency of biodegradation. The diversity of microbial species with the relevant metabolic capabilities is crucial for breaking down different components of hydrocarbons.

4. *Nutrient Availability*: Microbial biodegradation requires nutrients such as nitrogen, phosphorus, and trace elements. Adequate nutrient availability can enhance microbial activity and support the growth of hydrocarbon-degrading microorganisms. In some cases, adding nutrients (biostimulation) can accelerate biodegradation.

5. *Temperature*: Temperature plays a vital role in microbial activity. Biodegradation rates generally increase with higher temperatures, as microbial metabolism is typically more active in warmer environments. However, extreme temperatures can also inhibit microbial activity.

6. *pH Level*: The pH level of the contaminated environment affects the microbial community composition and enzyme activity. Different microorganisms have optimal pH ranges for their metabolic processes, and extreme pH levels can limit microbial growth and hydrocarbon degradation.

7. *Moisture Content*: Adequate moisture is essential for microbial activity and

the transport of hydrocarbons to the microorganisms. Insufficient moisture can limit biodegradation, particularly in arid environments.

8. *Presence of Inhibitors*: Some contaminants or environmental factors may inhibit microbial activity or the effectiveness of specific hydrocarbon-degrading enzymes. Heavy metals, certain chemical compounds, and extreme salinity are examples of inhibitors that can hinder bioremediation.

9. *Environmental Conditions*: The overall environmental conditions, including climate, weather, and ecosystem type, influence the overall success of bioremediation. For instance, hydrocarbons spilled into aquatic environments may have different biodegradation dynamics compared to those in terrestrial soils.

10. *Residence Time*: The length of time that hydrocarbons have been present in the environment (residence time) can also impact biodegradation. Longer exposure to hydrocarbons may result in the development of microbial communities adapted to the contaminants, leading to enhanced biodegradation.

Considering these factors and tailoring bioremediation strategies accordingly can significantly improve the efficiency and success of hydrocarbon biodegradation. However, it's essential to conduct site-specific assessments and monitoring to ensure the most appropriate and effective bioremediation approach is applied for each contaminated site.

Chapter - 9

Nanobioremediation

9.1. Introduction

In the current scenario, it has become a very challenging task to tackle or remediate various forms of environmental pollution by eco-friendly, economical and sustainable technologies. The area of study investigating the removal of contaminants from the environment is called environmental remediation and the methods used to remove or degrade the pollutants are called environmental remediation techniques. One of the most sought after technique for the environment restoration is bioremediation.

Nanobioremediation (NBR) is the use of nanoparticles/nanomaterials produced by plant, fungi, and/or bacteria by means of nanotechnology, for elimination of environmental contaminants such as heavy metals, organic and nonorganic waste. Nanobioremediation is an innovative and promising field at the intersection of nanotechnology and environmental biotechnology. It involves the use of nanomaterials, such as nanoparticles and nanoscale biologically active substances, to remediate and mitigate environmental pollution caused by various contaminants, including heavy metals, organic pollutants, and other hazardous substances. The traditional methods of environmental remediation often have limitations, such as high cost, long treatment times, and the generation of secondary pollutants. Nanobioremediation seeks to overcome these challenges by harnessing the unique properties of nanomaterials and the biological activities of microorganisms (Tripathi et al., 2022).

Applications of nanoparticles can be seen in almost every field of science like automobiles, cosmetics, agriculture, food, textiles, space, defense, engineering, medical fields, and environment. According to the US National Nanotechnology Initiative (NNI), nanotechnology is defined as "the understanding and control of matter at dimensions between approximately 1–100 nanometres, where unique

DOI: 10.1201/9781003787549-9

phenomena enable novel nanotechnology applications." Encompassing nanoscale science, engineering, and technology, nanotechnology involves imaging, measuring, modeling, and manipulating matter at this length scale. In the past few years, use of nanotechnology in contaminant removal has become prominent due to its small particle size, high surface area to volume ratio, easy injection to the site of action, flexibility for in situ and ex situ application, etc.

9.2. Difference between Nanoparticle and Nanomaterial

Nanoparticles and nanomaterials are related terms used in the field of nanotechnology, which is the science, engineering, and application of materials and structures at the nanoscale (typically at dimensions less than 100 nanometers). While the terms are often used interchangeably, there are some subtle distinctions between them:

Nanoparticles:

» Nanoparticles are submicroscopic particles with at least one dimension (length, width, or height) in the nanoscale range, typically between 1 and 100 nanometers.

» They can be composed of various materials, including metals (e.g., gold, silver, copper), metal oxides (e.g., titanium dioxide, iron oxide), semiconductors (e.g., quantum dots), polymers, and more.

» Nanoparticles can have unique properties and behaviors compared to their bulk counterparts due to their small size and high surface area-to-volume ratio. These properties make them valuable for a wide range of applications, including drug delivery, imaging, catalysis, and electronics.

Nanomaterials:

» Nanomaterials are a broader category of materials that encompass not only nanoparticles but also any material or structure that exhibits nanoscale dimensions or properties.

» Nanomaterials can include nanoparticles, nanowires, nanotubes, nanosheets, nanocomposites, and more. These materials may be composed of nanoparticles or may have a complex hierarchical structure at the nanoscale.

» The term "nanomaterial" is often used when discussing materials with diverse nanoscale features, which can have a wide range of properties and applications.

9.3. Shapes, Sizes, and Structures of Nanomaterials

In essence, nanotechnology uses particles with sizes between 1 and 100 nm to create functional systems that can be used to solve problems or carry out certain tasks. The unique size, shape, and structure of nanoparticles determine a variety of their features, including their reactivity, magnetism, stability, and optical qualities. Because of these qualities, nanoparticles can be used in a variety of industries, including food packaging, textiles, cosmetics, medicine delivery, and water filtration (Demir, 2021). The nanoparticles are referred to as nanospheres, nano-rods, nano-cubes, etc. based on their various shapes, which include rods, spheres, cubes, triangles, polygons, etc.

The dimensions of the nanomaterials can be used to organize their structure. The majority of nanomaterials are found in zero dimensions, such as fullerenes and atomic clusters; one dimension, such as nanofibers and nanowires; two dimensions, such as nanodisks and nanolayers; or three dimensions, such as embedded clusters.

9.4. Synthesis and Characterization

For the synthesis of nanoparticles, there are primarily two methods. There are two types of approaches: top-down and bottom-up. Top-down approaches, such as high energy ball milling, grinding, etching, laser pyrolysis, lithographic techniques, etc., break down larger systems to produce nanosized particles, while bottom-up approaches combine atoms to form clusters, which then aggregate to produce nanoparticles. Physical, chemical, and biological approaches can be used to create nanoparticles. Following synthesis, the nanoparticle must be characterized in order to determine its size, shape, surface charge, morphology, crystallographic nature, etc. Numerous methods, including scanning electron microscopy (SEM), transmission electron microscopy (TEM), X-ray diffraction (XRD), Fourier transform infrared spectroscopy (FTIR), scanning tunneling microscopy (STM), nuclear magnetic resonance (NMR), etc., can be used for this characterisation.

9.5. Types of Nanomaterials (NMs) and their Applications in Bioremediation and Biodegradation

Numerous NMs with various morphologies and forms are employed. Because of its huge surface area and reactivity, NMs are better adsorbents than other traditional tools, making them useful for treating industrial effluent and drinking water. Nanoscale zeolites, metal oxides, Pd/Fe0 bimetallic NPs, zinc oxide, carbon nanotubes, chitosan, and graphene nanosheets are examples of innovative nanoscale materials that have been studied for bioremediation over the years. Dendrimers, modified dendrimers, carbon nanotubes, calcium alginate, and multiwalled carbon nanotubes are among the various organic NMs that have been created for metal removal, radionucleotide remediation, hydrogenation of toluene, and degradation of phenol and crude oil. Similarly, As, U, nitrate, and PCBs have been remedied using TiO2 NPs, polysulfone-zVI, Ag-iron oxide/fly ash, poly (acrylic acid)-coated Fe oxide, and chitosan-gelatin/graphene oxide.

9.5.1. Metallic Nanoparticles

Metal NPs are purely made of the metals precursors. Metal-based nanoparticles are widely used in bioremediation processes due to their unique properties, such as high surface area, reactivity, and catalytic activity. These nanoparticles can be applied to remediate various types of contaminants, including heavy metals, organic pollutants, and emerging contaminants. Iron NPs are considered as one of the first nanoparticles to be used in environmental cleanup.

Role of metal-based nanoparticles in bioremediation are as follows:

i. **Zero-Valent Metals (NZVs):**

» *Zero-Valent Iron (nZVI)*: nZVI is one of the most commonly used nanoparticles in bioremediation. It is highly effective at reducing and immobilizing various heavy metal ions (e.g., Cr(VI), Pb(II)) and chlorinated organic compounds through redox reactions.

» *Zero-Valent Silver (nZVA)*: nZVA exhibits antimicrobial properties and can be used to inhibit the growth of bacteria, which can be beneficial in some bioremediation applications.

ii. **Metal Oxide Nanoparticles:**

» *Titanium Dioxide (TiO2)*: TiO2 nanoparticles are known for their photocatalytic activity. They can degrade organic contaminants when exposed to ultraviolet (UV) or visible light, making them useful for the removal of pollutants like dyes and pesticides.

» *Iron Oxide Nanoparticles (e.g., Fe3O4, Fe2O3)*: Iron oxide nanoparticles can be used both for adsorption of contaminants and for catalytic degradation of organic compounds. They are particularly effective for the removal of heavy metals and arsenic.

iii. Noble Metal Nanoparticles:

- *Gold Nanoparticles (AuNPs)*: AuNPs are often employed for their catalytic activity, which can enhance the degradation of organic pollutants in the presence of suitable reducing agents or enzymes.
- *Silver Nanoparticles (AgNPs)*: AgNPs not only have antimicrobial properties but can also catalyze the degradation of organic contaminants. They are used in conjunction with microorganisms in some bioremediation strategies.

iv. Cadmium Sulfide (CdS) Nanoparticles:

- CdS nanoparticles can be used to immobilize heavy metals like cadmium, lead, and zinc by forming stable metal sulfide complexes.

v. Metal Nanoparticle Composites:

- *Metal-Polymer Composites*: Combining metal nanoparticles with polymers can improve their stability and provide controlled release of the nanoparticles for targeted remediation.
- *Metal-Carbon Nanocomposites*: Integrating metal nanoparticles with carbon-based nanomaterials like carbon nanotubes or graphene can enhance adsorption and catalytic properties.

vi. Bimetallic Nanoparticles:

- *Bimetallic Nanoparticles (e.g., Pd-Pt, Fe-Ni)*: Bimetallic nanoparticles combine the reactivity of two different metals, allowing for enhanced catalytic activity and selectivity in degrading various contaminants.

vii. Magnetic Nanoparticles:

- *Magnetic Nanoparticles (MNPs)*: MNPs are often used for the removal of heavy metals and other contaminants from water by applying an external magnetic field, simplifying the recovery and reuse of the nanoparticles.

viii. Metal-Doped Nanoparticles:

- *Metal-Doped Nanoparticles* (e.g., Cu-doped ZnO): Incorporating metal ions into metal oxide nanoparticles can enhance their catalytic activity and selectivity for specific contaminants.

A detailed list of metallic NPs used for bioremediation is covered in Table 9.1.

Table 9.1. Role of various nanoparticle in bioremediation

Metallic NPs	Remediating moiety
Au NPs	4-Nitrophenol, Methylene blue, Tertiary dye effluent (methyl orange, acid orange 10, acid red 88)
AgNPs	Methylene blue, Textile effluent, 4-Nitrophenol, Coomassie brilliant blue G-250, Congo red, Organic dyes (methyl violet, eosin, safranin, methyl orange, methylene blue)
Cu NPs	Dichloromethane, Methylene blue, Methyl orange
Zn NPs	4-Chlorocatechol, Resorcinol, Fuchsine, Congo red and benzopurpurine 4B, Rhodamine B, Phenol, Cd(II), Rhodamine B, Methylene blue, Direct red 23, Eriochrome black-T dye, Brown CGG dye, Formaldehyde, Malachite green, Organic dyes
Fe NPs	Pentachlorophenol, Perchlorate, Chlorinated ethanes, Alachlor, pretilachlor, Ni(II), Brominated methanes, Dibenzo-p-dioxins and furans, Lindane, atrazine, Polychlorinated biphenyls (PCBs), Atrazine, Alachlor and atrazine, Pyrene, Cu(II), Cr(VI), As(V), Cr(VI), Cr(VI), Pb(II), hg(II), Metolachlor, Cu(II), Pb(II), Trichloroethylene, Cationic and anionic dyes, Nitrate, Chlorobenzene, Uranium, Dichloroethane, Cr(VI) and cd(II), Cd^{2+}, Dissolved sulfides

9.5.2. Carbon-Based Nanomaterials

Carbon-based nanomaterials are a class of materials composed primarily of carbon atoms arranged in nanoscale structures. They have gained considerable attention in the field of bioremediation due to their high surface area, strong adsorption properties, and potential for functionalization (Díez et al., 2021). These characteristics make them suitable for various applications in removing contaminants from the environment.

Carbon-based nanomaterials are used in bioremediation are as follows:

i. **Adsorption of Contaminants:**

   » *Heavy Metal Removal*: Carbon-based nanomaterials like activated carbon, graphene, and carbon nanotubes are effective adsorbents for heavy metals, including lead, cadmium, and mercury. They can be used to capture and immobilize these toxic ions from contaminated water and soil.

ii. Removal of Organic Pollutants:

» *Organic Compound Adsorption*: Carbon-based nanomaterials can adsorb organic pollutants such as polycyclic aromatic hydrocarbons (PAHs), pesticides, and volatile organic compounds (VOCs). They can be used to reduce the concentration of these contaminants in water and soil.

iii. Enhanced Biodegradation:

» *Carrier for Microbes*: Carbon-based nanomaterials can serve as carriers for beneficial microorganisms, such as bacteria or fungi, in bioremediation processes. They provide a protective environment for the microbes, enhance their survival, and promote the degradation of pollutants.

iv. Nanomaterial-Based Catalysts:

» *Catalytic Degradation*: Carbon-based nanomaterials, such as graphene and carbon nanotubes, can be functionalized or loaded with catalytic metals (e.g., palladium, platinum) to facilitate the breakdown of contaminants. These materials can serve as catalytic supports in advanced oxidation processes (AOPs) or in-situ chemical oxidation (ISCO) reactions.

v. Electrochemical Treatment:

» *Electrochemical Remediation*: Carbon-based nanomaterials can be used as electrodes in electrochemical remediation techniques. They can enhance the removal of various contaminants through processes like electrocoagulation, electrooxidation, and electro-Fenton reactions.

vi. Photocatalysis:

» *Photocatalytic Degradation*: Carbon-based nanomaterials, especially graphene and graphene oxide, can be modified to become photocatalysts. When exposed to light, they generate reactive species that can degrade organic contaminants in water and air.

vii. Nanocomposites:

» *Nanocomposite Materials*: Carbon-based nanomaterials can be integrated into composite materials with other substances, such as polymers or metal nanoparticles, to create multifunctional materials for targeted remediation of specific contaminants.

viii. Environmental Sensing:

- *Contaminant Monitoring*: Carbon-based nanomaterials are used as sensors and detectors for monitoring environmental contaminants. They can provide real-time data on pollutant concentrations and help in assessing the effectiveness of bioremediation processes.

ix. **Groundwater and Soil Remediation**:

- *In-situ Remediation*: Carbon-based nanomaterials can be injected or applied in-situ to contaminated groundwater and soil to facilitate the remediation of subsurface contaminants.

x. **Wastewater Treatment:**

- *Advanced Wastewater Treatment*: Carbon-based nanomaterials are used in wastewater treatment processes to remove a wide range of contaminants, including dyes, pharmaceuticals, and emerging contaminants.

9.5.3. Polymeric Nanomaterials

Polymeric nanoparticles are organic-based nanoparticles. Depending upon the preparation method, these types of nanoparticles have structures shaped like nanocapsules or nanospheres. Polymeric nanomaterials play a significant role in bioremediation processes, offering various advantages such as controlled release, increased stability, and improved biocompatibility (Han et al., 2018). These materials are used in a variety of ways to enhance the efficiency of bioremediation and address environmental contamination issues.

Applications of polymeric nanomaterials in bioremediation are as follows:

i. **Encapsulation of Microorganisms:**

- Polymeric nanoparticles can encapsulate beneficial microorganisms such as bacteria or fungi, protecting them from harsh environmental conditions and ensuring controlled release at the contaminated site. This approach enhances the survival and activity of the microorganisms in bioremediation processes.

ii. **Controlled Release of Nutrients and Biostimulants:**

- Polymers can be used to encapsulate nutrients, biostimulants, or growth-promoting substances, which are slowly released into the environment to support the growth and metabolic activity of indigenous or introduced microorganisms involved in bioremediation.

iii. **Sorption and Adsorption of Contaminants:**

» Polymeric nanomaterials, such as functionalized polymers and hydrogels, can adsorb or sorb various contaminants, including heavy metals, organic pollutants, and nutrients. This property can be harnessed to remove or immobilize contaminants from water, soil, or sediments.

iv. Enhanced Transport and Mobility:

» Polymeric nanoparticles can improve the transport and mobility of microorganisms and nutrients within the contaminated site. They can help distribute bioremediation agents more effectively, ensuring a higher contact surface area with contaminants.

v. Biodegradable Polymers:

» Biodegradable polymers are designed to degrade over time, leaving behind no harmful residues. They can be used as carriers for bioremediation agents and gradually break down, releasing the agents into the environment while minimizing environmental impact.

vi. Functionalized Polymers:

» Functionalized polymers with specific chemical groups can be tailored to target and bind specific contaminants selectively. These polymers can be used to remove or recover contaminants such as heavy metals or organic pollutants.

vii. Hydrogels:

» Hydrogels are water-absorbent polymers that can retain large volumes of water. They are used to create moisture-rich environments in soil or sediments, promoting microbial activity and the biodegradation of contaminants.

viii. Nanocomposite Materials:

» Polymeric nanomaterials can be combined with other nanomaterials, such as carbon-based nanomaterials or metal nanoparticles, to create multifunctional nanocomposites with enhanced adsorption, catalytic, or transport properties for bioremediation applications.

ix. pH and Temperature-Controlled Release:

» Some polymeric nanomaterials can be designed to release their payloads in response to changes in pH or temperature, allowing for tailored and controlled release of bioremediation agents in specific environmental conditions.

x. Wastewater Treatment:

» Polymeric nanomaterials are used in wastewater treatment to remove contaminants, including heavy metals, organic pollutants, and pathogens, through processes like adsorption, coagulation, and flocculation.

9.5.4. Hybrid nanomaterials

Hybrid nanomaterials are composite materials consisting of two or more different types of nanomaterials or nanoparticles combined at the nanoscale (Laurila et al., 2017).

Roles of Hybrid Nanomaterials in Bioremediation:

i. **Enhanced Adsorption:**

» Hybrid nanomaterials can provide a higher surface area and tailored surface chemistry, resulting in improved adsorption of contaminants, including heavy metals, organic pollutants, and dyes. They can be used to remove pollutants from water and soil effectively.

ii. **Selective Targeting:**

» Hybrid nanomaterials can be designed to target specific contaminants or groups of contaminants. This selectivity allows for the efficient removal of particular pollutants while leaving others untouched, reducing the overall environmental impact.

iii. **Controlled Release of Bioremediation Agents:**

» Hybrid nanomaterials can serve as carriers for bioremediation agents, such as enzymes or beneficial microorganisms. They can control the release of these agents into the environment, ensuring sustained and targeted remediation.

iv. **Catalytic Degradation:**

» Certain hybrid nanomaterials, especially those incorporating metal nanoparticles, exhibit catalytic activity. This property can be harnessed to facilitate the breakdown of organic pollutants or enhance the redox reactions involved in contaminant removal processes.

v. **Environmental Remediation in Complex Matrices:**

» Hybrid nanomaterials are effective in addressing complex and challenging environmental matrices, such as groundwater or soil, where contaminants may exist in diverse forms and concentrations. Their versatility allows them to adapt to various remediation scenarios.

9.5.5. Ceramic-Based NPs

These are made up of oxides, carbonates, and phosphates. They show good resistance to chemicals and heat. In bioremediation, ceramic-based nanoparticles have found applications in addressing various environmental contamination challenges (Adhikari et al., 2018).

i. Adsorption of Contaminants:

» Ceramic-based nanoparticles, such as titanium dioxide (TiO_2), zirconium oxide (ZrO_2), and aluminum oxide (Al2O3), have high surface areas and can adsorb a wide range of contaminants, including heavy metals, dyes, and organic pollutants, from aqueous solutions and soil. Their adsorption properties are valuable for the removal and immobilization of contaminants.

ii. Catalytic Degradation:

» Some ceramic-based nanoparticles, particularly TiO2 and ZrO2, exhibit photocatalytic properties. When exposed to ultraviolet (UV) or visible light, these nanoparticles can generate reactive oxygen species (ROS) and initiate degradation processes for organic pollutants. This photocatalytic activity is useful in degrading contaminants in water and air.

iii. Redox Reactions:

» Certain ceramic nanoparticles, like zero-valent iron (nZVI), are capable of participating in redox reactions with contaminants. nZVI can reduce and immobilize heavy metals, metalloids, and chlorinated organic compounds through redox reactions, making it effective in groundwater and soil remediation.

iv. Stabilization of Contaminants:

» Ceramic-based nanoparticles can stabilize contaminants in soil and sediment by encapsulating or binding them, preventing their migration and reducing their bioavailability to organisms. This is particularly useful for controlling the spread of contaminants in the environment.

v. pH Adjustment

» Ceramic-based nanoparticles can be used to adjust the pH of contaminated environments. For instance, Al2O3 nanoparticles can act as pH modifiers, helping to maintain optimal pH conditions for bioremediation processes.

vi. Nanocomposite Materials

» Ceramic nanoparticles can be integrated into nanocomposite materials, such as ceramic-polymer composites, to enhance their adsorption and catalytic properties. These nanocomposites can be tailored for specific bioremediation applications.

vii. Groundwater and Soil Remediation

» Ceramic-based nanoparticles can be injected or applied in-situ to contaminated groundwater and soil environments to facilitate the remediation of subsurface contaminants. They can improve the transport and distribution of bioremediation agents or enhance the immobilization of pollutants.

viii. Immobilization of Microorganisms and Enzymes:

» Ceramic nanoparticles can immobilize microorganisms or enzymes for bioremediation purposes. They provide a stable matrix for microbial growth and enzyme activity, enhancing the degradation of contaminants.

ix. Wastewater Treatment:

» Ceramic-based nanoparticles are used in wastewater treatment processes to remove contaminants, including heavy metals, dyes, and organic compounds. They can serve as adsorbents, coagulants, or catalysts in various treatment systems.

9.5.6. Lipid Based Nanomaterials

Lipid nanoparticles are generally spherical, with a diameter ranging from 10 to 100nm. Lipid nanoparticles (LNPs) are a class of nanoscale materials composed primarily of lipids, which are organic molecules such as fats and phospholipids. LNPs have gained attention in bioremediation due to their biocompatibility, versatility, and ability to encapsulate and deliver various bioremediation agents.

i. Encapsulation of Bioremediation Agents:

» LNPs can encapsulate and protect bioremediation agents such as enzymes, microorganisms, or biodegradable polymers. This encapsulation helps shield the agents from harsh environmental conditions and delivers them to the contaminated site effectively.

ii. Controlled Release of Bioremediation Agents:

» LNPs can be designed to release their payload of bioremediation agents gradually over time. This controlled release ensures a sustained presence of active agents at the contaminated site, optimizing remediation effectiveness.

iii. Targeted Delivery:

» LNPs can be surface-modified or functionalized to target specific contaminants or contaminated areas. This targeted delivery minimizes the dispersion of bioremediation agents in non-contaminated areas and reduces ecological impact.

iv. Enhanced Stability:

» LNPs enhance the stability of bioremediation agents, protecting them from degradation by temperature, pH extremes, or exposure to toxic substances commonly found in contaminated environments.

v. Protection Against Predation:

» LNPs can protect encapsulated microorganisms from predation by protozoa and other microorganisms present in the contaminated site, allowing for more efficient pollutant degradation.

vi. Immobilization of Microorganisms and Enzymes:

» LNPs can immobilize microorganisms or enzymes on their surfaces, providing a stable matrix for microbial growth and enzyme activity. This immobilization enhances the ability of bioremediation agents to degrade contaminants.

vii. Soil and Sediment Remediation:

» LNPs can be introduced into contaminated soils or sediments, facilitating the in-situ remediation of various pollutants. They enhance the bioavailability of contaminants to microorganisms and enzymes, improving remediation efficiency.

viii. Groundwater Remediation:

» LNPs can be used for groundwater remediation efforts to deliver bioremediation agents to subsurface contamination plumes. Their controlled release properties can help maintain a stable concentration of active agents over time.

ix. Enhanced Biodegradation:

» LNPs support the growth and metabolic activity of indigenous or introduced microorganisms. They create a favorable microenvironment for microbial communities, promoting the biodegradation of contaminants.

x. Complex Contaminant Matrices:

» LNPs are effective in addressing complex environmental matrices, such as groundwater, soil, or sediments, where contaminants may exist in diverse forms and concentrations.

9.6. Key components of Nanobioremediation include:

1. Nanomaterials:

Nanomaterials have shown significant potential in the field of bioremediation, which is the process of using biological organisms or their derivatives to remove or neutralize pollutants from the environment (Wu et al., 2020). Nanomaterials can enhance the efficiency and effectiveness of bioremediation processes in several ways:

» *Increased Surface Area*: Nanomaterials have a high surface area-to-volume ratio due to their small size, which allows them to adsorb or immobilize contaminants more effectively. This property is particularly useful for the removal of heavy metals and organic pollutants from soil and water.

» *Enhanced Sorption*: Nanomaterials such as carbon nanotubes, graphene, and various nanoscale metals (e.g., iron nanoparticles) can adsorb contaminants onto their surfaces, making it easier for microorganisms to degrade or transform the pollutants. These materials can act as carriers or supports for bioremediation agents.

» *Improved Transport and Delivery*: Nanoscale materials can improve the transport and delivery of bioremediation agents to the target sites. For example, nanoparticles can be used to encapsulate and protect sensitive microorganisms or enzymes and deliver them to contaminated areas.

» *Catalytic Activity*: Certain nanomaterials, like zero-valent iron nanoparticles, can act as catalysts to facilitate the degradation of contaminants. They can enhance the efficiency of microbial processes by providing a surface for reactions to occur or by generating reactive species that help break down pollutants.

» *Monitoring and Detection*: Nanosensors and nanoprobes can be employed to monitor the progress of bioremediation processes. These nanomaterial-based tools can provide real-time data on pollutant concentrations and microbial activity, allowing for better control and optimization of bioremediation strategies.

» *Reduction of Toxicity*: Some nanomaterials have the potential to reduce the toxicity of contaminants. For example, certain nanoparticles can transform toxic heavy metal ions into less harmful forms or sequester them, preventing their mobility in the environment.

» *Selective Targeting*: Nanomaterials can be engineered to selectively target specific contaminants, which can increase the efficiency of bioremediation while minimizing the impact on non-targeted species.

However, it's essential to consider the potential environmental and health risks associated with the use of nanomaterials in bioremediation. The release of nanoparticles into the environment can have unintended consequences, and their long-term behavior and effects are still a subject of research and concern.

2. Bioremediation Agents:

Nanobioremediation often involves the use of microorganisms (bacteria, fungi, or algae) or their enzymes to biodegrade or transform contaminants. These bioremediation agents play a crucial role in breaking down or immobilizing pollutants.

3. Surface Modification:

Nanomaterials can be surface-modified or functionalized to improve their interaction with contaminants and bioremediation agents. Surface modifications can enhance the adsorption, catalytic activity, and selectivity of nanomaterials.

4. Contaminant-Specific Design:

Nanobioremediation strategies are often tailored to specific contaminants. The choice of nanomaterials and bioremediation agents can vary depending on the type of pollutant to be remediated (e.g., heavy metals, organic compounds, pesticides).

5. Delivery Systems:

Nanomaterials can be designed as carriers to deliver bioremediation agents to the contaminated sites effectively. This ensures that the agents reach the target pollutants in sufficient quantities.

6. Monitoring and Characterization Tools:

To assess the progress and effectiveness of nanobioremediation, various monitoring and characterization tools are essential. These may include nanosensors, spectroscopic techniques, and analytical chemistry methods for tracking pollutant concentrations and microbial activity.

7. Environmental Safety and Regulations:

Due to the potential risks associated with the use of nanoparticles in the environment, it's crucial to consider safety and regulatory aspects. Environmental impact assessments and risk evaluations should be part of the nanobioremediation process.

8. Site Assessment and Planning:

An initial site assessment is critical to understand the nature and extent of contamination. Site-specific conditions, such as soil composition, groundwater flow, and the presence of native microorganisms, should be considered when designing nanobioremediation strategies.

9. Integration with Traditional Remediation Techniques:

Nanobioremediation can be used in conjunction with traditional remediation methods, such as soil excavation, to achieve more comprehensive and effective cleanup of contaminated sites.

10. Long-Term Monitoring and Maintenance:

After implementing nanobioremediation, ongoing monitoring and maintenance are necessary to ensure that the remediation objectives are met and that there are no adverse environmental impacts.

11. Public and Stakeholder Engagement:

Effective communication with the public and stakeholders is important to address concerns, provide information on the remediation process, and gain support for nanobioremediation projects.

9.7. The process of Nanobioremediation

(i) Site Assessment:

The first step is to conduct a thorough assessment of the contaminated site. This includes identifying the type and extent of contamination, the nature of pollutants present, and the specific environmental conditions at the site. This information helps in designing an effective nanobioremediation strategy.

- *Characterization of Contaminants*: Identify and characterize the contaminants present at the site. This includes determining the type of pollutants (e.g., heavy metals, organic chemicals, petroleum hydrocarbons), their concentrations, distribution, and chemical properties. Understanding the contaminants is essential for selecting appropriate nanomaterials and microorganisms.

- *Site History and Land Use:* Gather historical information about the site, including its previous land use, industrial activities, and any past remediation efforts. This context can help in understanding the sources and extent of contamination.

- *Site Geology and Hydrogeology:* Assess the geological and hydrogeological conditions of the site. This involves studying the soil types, groundwater flow patterns, and subsurface characteristics. Knowledge of site hydrogeology is critical for understanding how contaminants may have migrated and how nanomaterials and microorganisms can be effectively delivered to the contaminated zones.

- *Ecological Assessment:* Evaluate the ecological impact of the contamination on the site's ecosystems. Assess the health and diversity of plant and animal species in the affected area. This information helps in understanding the potential collateral effects of nanobioremediation and guides restoration efforts.

- *Regulatory Requirements*: Identify and understand the applicable environmental regulations and guidelines governing the site's cleanup. Compliance with these regulations is essential throughout the remediation process.

- *Human Health and Safety Concerns*: Assess any potential risks to human health and safety associated with the site's contamination. Determine if there are nearby communities or sensitive receptors that may be affected.

- *Stakeholder Engagement*: Engage with relevant stakeholders, including local communities, regulatory agencies, and site owners, to gather input and address concerns. Open communication is important for building trust and ensuring a successful remediation process.

- *Sampling and Data Collection*: Conduct field sampling and laboratory analysis to obtain representative data on contaminant concentrations and distribution. Sampling should be performed at different depths and locations within the site to create a comprehensive contamination profile.

- *Data Integration and Analysis*: Analyze and integrate the collected data to develop a clear understanding of the site's contamination issues. Use tools such as geographic information systems (GIS) to create contamination maps and models.

» *Risk Assessment*: Perform a risk assessment to evaluate the potential environmental and human health risks associated with the contamination. This assessment informs the remediation goals and priorities.

» *Feasibility Study*: Conduct a feasibility study to determine if nanobioremediation is a viable and cost-effective remediation option based on the site-specific conditions and contaminant characteristics.

(ii) Selection of Nanomaterials:

The selection of nanomaterials for nanobioremediation is a critical step that involves choosing nanoparticles with properties that are well-suited to the specific contaminants and environmental conditions at the contaminated site. Here is a process for selecting nanomaterials for nanobioremediation:

» *Identify Contaminants*: Begin by identifying the types of contaminants present at the contaminated site. This includes understanding whether the contaminants are heavy metals, organic compounds, pesticides, hydrocarbons, or other substances. Different nanomaterials are effective for different types of contaminants.

» *Understand Contaminant Properties:* Gain a thorough understanding of the chemical and physical properties of the contaminants. This includes factors such as solubility, reactivity, toxicity, and mobility. The nanomaterials selected should be capable of interacting with or immobilizing these contaminants effectively.

» *Review Nanomaterial Options:* Research and review the available nanomaterial options that have shown promise in previous nanobioremediation studies. Common nanomaterials used in nanobioremediation include:

- Zero-valent metals (e.g., zero-valent iron, zero-valent copper)
- Metal oxide nanoparticles (e.g., titanium dioxide, iron oxide)
- Carbon-based nanoparticles (e.g., carbon nanotubes, graphene oxide)
- Clay nanoparticles (e.g., montmorillonite)
- Nanoscale porous materials (e.g., zeolites)

» *Consider Nanomaterial Properties:* Evaluate the properties of each potential nanomaterial, including:

- Surface area: Higher surface area allows for more efficient adsorption and interaction with contaminants.
- Surface charge: Surface charge can influence the affinity of nanomaterials for specific contaminants.
- Reactivity: Some nanomaterials may be more reactive and capable of catalyzing reactions with contaminants.
- Stability: Assess the stability of nanomaterials under the environmental conditions of the site.
- Biocompatibility: Consider whether the nanomaterial is biocompatible with the selected microorganisms.

» *Surface Modification:* Consider whether the selected nanomaterials require surface modification to enhance their effectiveness. Surface modifications can include functionalization with organic molecules or the addition of specific functional groups to increase their reactivity or stability.

» *Compatibility with Microorganisms:* Ensure that the chosen nanomaterials are compatible with the selected microorganisms. Nanomaterials should not inhibit microbial growth or activity but should instead work synergistically with the microorganisms.

» *Evaluate Cost and Availability:* Consider the cost and availability of the chosen nanomaterials. Some nanomaterials may be expensive or difficult to obtain in large quantities, which can impact the feasibility of the remediation project.

» *Pilot Studies:* Conduct small-scale pilot studies or laboratory experiments to test the effectiveness of selected nanomaterials in removing or immobilizing the contaminants of concern. These studies can help validate the suitability of the chosen nanomaterials before full-scale implementation.

» *Risk Assessment:* Perform a risk assessment to identify any potential risks associated with the use of nanomaterials at the specific site. Address any concerns related to the environmental impact of the nanomaterials.

» *Finalize Selection:* Based on the evaluation and testing, finalize the selection of nanomaterials that are most suitable for the nanobioremediation project. Ensure that the chosen nanomaterials align with the project goals and regulatory requirements.

The selection of nanomaterials for nanobioremediation should be a well-informed and systematic process, taking into account the unique characteristics of the contaminants and the site-specific conditions. Additionally, ongoing monitoring during the remediation process can help assess the performance of the chosen nanomaterials and make any necessary adjustments.

(iii) Modification of Nanomaterials:

Modify the selected nanomaterials to enhance their effectiveness in the remediation process. Surface modifications or functionalization can improve their stability, dispersibility, and affinity for the contaminants.

(iv) Selection of Bioremediation Agents:

Choose suitable microorganisms, such as bacteria, fungi, or algae, that are capable of metabolizing or interacting with the specific contaminants present at the site. These microorganisms should be able to thrive in the site's environmental conditions.

(v) Nanomaterial-Microorganism Interaction:

The process of nanomaterial-microorganism interaction involves various stages, from the initial contact between nanomaterials and microorganisms to potential effects on microorganisms and vice versa. Below is a generalized process describing the key steps in this interaction:

» *Contact and Dispersion*: Nanomaterials are introduced into the environment or biological system where microorganisms are present. These nanomaterials may be in the form of nanoparticles, nanotubes, or other nanoscale structures. The initial step is physical contact between the nanomaterials and microorganisms.

» *Surface Interaction:* Nanomaterials have specific surface properties, including charge, hydrophobicity, and functional groups, that influence their interaction with microorganisms. The surface properties can affect the attachment of microorganisms to nanomaterials.

» *Adhesion and Biofilm Formation:* Microorganisms may adhere to the surfaces of nanomaterials, forming biofilms. Biofilm formation is a common microbial response to surfaces and can involve the secretion of extracellular polymeric substances (EPS) that help microorganisms adhere to and colonize the nanomaterial surfaces.

- *Uptake by Microorganisms:* Depending on their size and properties, some nanoparticles can be internalized by microorganisms through processes like endocytosis or passive diffusion. This internalization can vary among different microorganisms and nanomaterials.
- *Intracellular Effects:* Once internalized, nanomaterials may interact with cellular components of microorganisms, potentially affecting cellular functions. This can include interfering with metabolic pathways, disrupting cell membranes, or affecting the integrity of genetic material.
- *Toxicity and Stress Responses:* Nanomaterials can induce stress responses in microorganisms, leading to the activation of defense mechanisms. The response may include the production of reactive oxygen species (ROS) and the upregulation of stress-related genes.
- *Microbial Adaptation and Resistance:* Prolonged exposure to nanomaterials may lead to microbial adaptation and the development of resistance mechanisms. Some microorganisms can evolve to tolerate or detoxify nanomaterials, reducing their effectiveness.
- *Release of Ions or Chemicals:* Certain nanomaterials, like silver nanoparticles, may release ions or chemicals that can have antimicrobial properties. These released ions can interact with microbial cells and disrupt their functions.
- *Environmental Fate and Transport:* In natural environments, the fate and transport of nanomaterials can influence their interaction with microorganisms. Factors such as sedimentation, aggregation, and chemical transformations can affect nanomaterial bioavailability to microorganisms.
- *Effects on Nanomaterials:* Microorganisms can also influence nanomaterial behavior. For example, microbial activities, such as secretion of EPS, can alter the surface chemistry and aggregation of nanomaterials.
- *Biodegradation and Transformation:* Some microorganisms possess the ability to degrade or transform certain nanomaterials. This can lead to changes in nanomaterial properties and may affect their environmental persistence.
- *Overall Impact:* The interaction between nanomaterials and microorganisms can have various outcomes, including antimicrobial effects, biodegradation, altered microbial community structure, or potential ecological implications.

(vi) Monitoring and Control:

Continuously monitor the progress of the nanobioremediation process. Parameters such as contaminant concentrations, microbial activity, and nanomaterial behavior should be regularly assessed to ensure the effectiveness of the remediation effort. Adjustments can be made as necessary.

(vii) Biodegradation and Transformation:

Microorganisms metabolize and degrade the contaminants, converting them into less harmful or non-toxic substances. This biodegradation process is often enhanced by the presence of nanomaterials, which can increase the surface area for microbial attachment or improve the bioavailability of contaminants.

(viii) Remediation Completion:

Once the targeted contaminants have been sufficiently reduced to meet regulatory standards or environmental goals, the nanobioremediation process can be considered complete. Post-remediation monitoring may still be necessary to ensure long-term effectiveness and ecological recovery.

(ix) Site Restoration:

After successful remediation, the site can undergo restoration to return it to its natural or intended state. This may involve re-vegetation, habitat reconstruction, or other ecological restoration efforts.

(x) Regulatory Compliance:

Ensure that the remediation process complies with all relevant environmental regulations and standards. Documentation and reporting may be required to demonstrate that the site has been effectively cleaned up.

The use of various bioagent for nanobioremediation of environmental contaminants is shown in Table 9.2.

Table 9.2: Remediation of environmental contaminants using nano-bioremediation

Nanoparticle	Bioagent	Contaminant	Remark
Carbon nanotubes	Shewanella oneidensis	Cr(VI)	The MR-1 strain immobilized by CNT infused CA beads could remove four times higher Cr (VI) than the free cells or CNTs or CA beads
nZVI	*Dehalococcoides spp.*	TCE	This study showed that nZVI stimulated the metabolic activity of methanogens but deactivated the dechlorinating bacteria, but after a lag phase the dechlorinating bacteria could again remove TCE producing ethene as by product
Pd(0) nanoparticles	Shewanella oneidensis	PCBs	The bio-Pd formed from the microbial reduction effectively dechlorinated around 90% of PCBs producing less toxic by-products
Fe3O4 nanoparticles	Sphingomonas sp.	Carbazole	The Fe_3O_4 nanoparticles bound to the surface of the bacterial strain showed no increased degradation than the free cells but showed amazing reusability. Another advantage of using magnetic nanoparticles is it can be separated from the microorganism using an external magnet source
Magnetic Fe3O4 nanoparticles	*Pseudomonas delafieldii*	Dibenzothiophene	The magnetic nanoparticle coated microbial cells showed greater biodesulfurization of dibenzothiophene than the free cells or cells coated with celite. It is also observed that it can be reused more than five times

nZVI	*Paracoccus sp. strain*	Nitrate	Lower conc. of nZVI (50 mg/L) enhanced denitrification process along with slight microbial toxicity, while higher conc. (1000 mg/L) significantly reduced denitrification rate
Fe3O4 nanoparticles/ gellan gum gel beads	*Sphingomonas sp. strain*	Carbazole	The microbial cells immobilized in Fe_3O_4 nanoparticles/gellan gum gel beads degraded higher carbazole than the free cells and the non-magnetically immobilized cells. This integrated system showed progressive increase in degradation when being recycled
Pd/nFe	Laccase derived from *Trametes versicolor*	Triclosan	The remediation of triclosan was solely achieved by Fe nanoparticles. However the degraded by-products were further converted to nontoxic compounds by the laccase secreted from *T. versicolor* strain
Bio-Pd nanoparticle	*C. pasteurianum BC1*	Cr(VI)	C. pasteurianum reduced the Pd(II) ions to Pd nanoparticles which stayed in the form of bio-Pd in the cell membrane and cytoplasm of the organism. It successfully catalyzed the Cr(VI) reduction reaction and also produced hydrogen gas
Pd/nFe	*Sphingomonas wittichii* RW1 (DSM 6014)	2,3,7,8-tetrachloro dibenzop-dioxin (2,3,7,8-TeCDD	The highly toxic dioxin isomer is recalcitrant in nature and its degradation could not be acquired easily through a single technique. The degradation was accomplished by using the Pd/nFe nanoparticles and the Sphingomonas strain sequentially

Pd/nFe	Burkholderia xenovorans	Polychlorinated biphenyl (PCB) Aroclor 1248	Pd/nFe nanoparticles efficiently dechlorinated the bi-, tri-, tetra-, penta-, hexa-chlorinated biphenyls Into biodegradable intermediates which were then easily degraded by Burkholderia xenovorans
nZVI-C-A beads	*Bacillus subtilis, E. coli, and Acinetobacter junii*	Cr(VI)	The thin biofilm covering the nZVI entrapped calcium alginate beads removed around 92% of Cr (VI) showing enhanced removal by the combined technology

Chapter - 10

Factors Affecting Bioremediation of Environmental Pollutants

10.1. Introduction

Bioremediation, as a natural and sustainable method for removal of environmental pollutants, has gained significant attention in recent years due to its potential to combat pollution effectively without causing any adverse effect on the environment with expense of biological activities. Bioremediation means using biological agents to clean environment with expense of biological activities. Indiscriminate utilization of natural resources and industrial growth increased the various toxic substances in the environment. Increase in the pollution has lead to increase in toxic substances in the environment and being referred to as most effective management tool bioremediation has tremendous future to be called as Eco biotechnology. Hence we can infer that bioremediation is a attractive tool used at number of sites which were degraded and attained their original position with onset of this technology. Bioremediation technology uses the microbes to remediate contaminated environment and brings back it to original position. Bioremediation has also been a solution for various emerging problems (Srivastava et al., 2014).

Bioremediation is a process that utilizes living organisms, such as bacteria, fungi, and plants, to break down or transform contaminants into less harmful substances, ultimately restoring the environmental quality. This approach has garnered attention due to its cost-effectiveness, minimal environmental impact, and potential for long-term sustainability. However, the success of bioremediation strategies is inherently tied to various factors that must be considered and managed to achieve optimal results. One of the primary factors affecting bioremediation success is the type and nature of the contaminant. Different pollutants require specific microbial communities or plant species for effective degradation. For instance, hydrocarbon-degrading bacteria are essential for remediating oil spills, while heavy metal-contaminated

DOI: 10.1201/9781003787549-10

sites may require hyperaccumulating plants to extract and immobilize the metals. Understanding the biochemical pathways and metabolic capabilities of the selected bioremediation agents is crucial in tailoring strategies to the contaminant at hand.

Environmental conditions also play a pivotal role in bioremediation outcomes. Temperature, pH, moisture content, and oxygen availability can significantly impact the activity and growth of microorganisms involved in remediation processes. Suboptimal conditions may hinder microbial activity, delaying the remediation process or rendering it ineffective. Therefore, careful monitoring and manipulation of environmental parameters are essential to ensure favorable conditions for the bioremediation process. The presence of co-contaminants represents another important factor influencing bioremediation. In many polluted sites, contaminants are rarely isolated, and the interactions between different pollutants can be complex. Some compounds may inhibit or enhance the degradation of others. Thus, a comprehensive understanding of the pollutant mixture and its potential synergistic or antagonistic effects on bioremediation is crucial for successful implementation.

The choice of bioremediation strategy is also dependent on the site-specific conditions and regulatory requirements. In situ bioremediation involves treating contaminants on-site without excavation or transport, while ex situ methods involve the removal of contaminated material to be treated elsewhere. Both approaches have their advantages and limitations, and selecting the most suitable strategy requires a thorough site assessment and consideration of cost, time, and logistical constraints. Bioremediation can be significantly affected by the indigenous microbial population at a contaminated site. Indigenous microorganisms may compete with introduced bioremediation agents for resources or exhibit antagonistic behaviors that hinder successful remediation. Understanding the microbial ecology of the site and possibly enhancing the population of beneficial microorganisms through bioaugmentation are essential strategies for overcoming this challenge.

The regulatory and public acceptance aspects of bioremediation should not be underestimated. Compliance with local and national regulations, as well as public perception and involvement, can influence the feasibility and success of bioremediation projects. Effective communication and transparency with stakeholders, along with adherence to regulatory guidelines, are vital for gaining support and ensuring the long-term sustainability of bioremediation efforts. Understanding the nature of the contaminants, optimizing environmental conditions, addressing co-contaminants, selecting appropriate strategies, managing indigenous microbial populations, and navigating regulatory and public acceptance challenges are all crucial elements in achieving effective and sustainable bioremediation solutions. The complex interplay of these factors underscores the need for a multidisciplinary and holistic approach to environmental cleanup, recognizing that no single solution fits all scenarios.

10.2. Factors effecting Bioremediation

The principle of bioremediation is that microorganisms (mostly bacteria or fungi) are used to degrade hazardous contaminants or covert them to less harmful forms. Thus, bioremediation of contaminants is an application of the microbial metabolic activity. Microorganisms, with their enzymatic pathways, act as biocatalysts and facilitate the progress of biochemical reactions that detoxify the targeted contaminants. As a result, bioremediation processes are only applicable in environments that can sustain life. The microbes act upon the contaminants only when they have access to a variety of materials-compounds to help them extract nutrients and energy to build more cells. In very few cases the natural conditions that exist at the contaminated site provide all the essential materials in large enough amount that bioremediation can occur without human intervention - a process called intrinsic bioremediation. Frequently, bioremediation needs the construction of engineered systems to supply microbe stimulating materials - a process called engineered bioremediation. Engineered bioremediation purely depends on accelerating the desired biodegradation reactions by encouraging the growth of more organisms, as well as by optimizing the environment in which the organisms must carry out the detoxification reactions.

The metabolic characteristics of the microorganisms in association with the physicochemical properties of the object contaminants determine whether a specific microorganism - contaminant interaction is possible. The actual successful interaction between the two, however, depends on the environmental conditions of the site of the interaction. Specific constrains should therefore be fulfilled for a successful bioremediation attempt. These constrains encompass the microbial, chemical and environmental characteristics of the targeted site.

10.2.1. Microbial constrains

Microbial constraints in bioremediation indicate to the different factors that regulate the activity and effectiveness of microorganisms involved in the degradation of any pollutants either in ex-situ or in-situ bioremediation process. These constraints can effect the ultimate performance of bioremediation process. To understand the different constrains of microbial community during growth in presence of any pollutants for bioremediation is important step for optimization of bioremediation strategies (Mishra et al., 2021). An optimum bioremediation results depend on utilization of the specific microorganisms responsible for utilization of pollutants as nutrient.

Such microbial populations can be inform of mixed population which may be also called as consortium isolated from the polluted site or naturally occurring species or genetically modified microorganisms. This knowledge gap is not necessarily the result of a scientific inability but rather of the continuous dynamic adaptation of the microbial species to their environments. At several occasions the naturally occurring microorganisms are not correctly identified, this created a gap between the knowledge of microorganism and their performance. In general most of microorganism remains very versatile for their nutritional adaptability in nature. These adaptational mechanisms advance through selection processes in which variant species with a specific survival advantage for the given environment take over and survive successfully. The survival advantage often relies on the ability of an organism to metabolise as substrate organic molecules (pollutants) existing in a given site. Modern microbiological techniques resolve the identification issue or problem of such transconjugants that originate from a background microbial population confirming that such processes are active in bioremediation practice. Horizontal transfer of catabolic plasmids among different species existing within a site may also result into species that possess enhanced catabolic or resistance potential. Such plasmid containing bacteria have been separated from polluted sites. Species that can through such plasmid transfer catabolise as single carbon source hazardous xenobiotics (as for example 3-chlorobenzoate).

Genetically modified microorganisms (GMOs) have often been presented as offering a major potential advantage for bioremediation. The development of recombinant DNA and other genetic engineering technologies, in the late 1970s, was believed that could be widely applied for environmentally-beneficial purposes, including the clean-up of contaminated soil and water. The continuously growing knowledge on catabolic pathways and critical enzymes provides the basis for the rational genetic design of new and improved enzymes and pathways for the development effective processes. Many researchers had expected that genetically modified organisms having novel biochemical traits or enzymatic activities would quickly find broad applicability in bioremediation of hazardous chemicals from the environment. However the practical impact of GMOs is likely to remain low for many key reasons. Public, economic and technical issues associated with the let go of genetically engineered, or recombinant, microbial species into an open environment usually arise. Many site owners, consultants and regulators are more comfortable choosing technologies and methods with which they are familiar, have a long track record of success and thus a greater predictability. Legislative reasons are associated with the strict control on the release of such organisms into the environment. There is significant concern about the long term survival of genetically engineered species into a natural environment where they would have to compete with the naturally existing consortia that had ample time to adapt to the prevailing environmental conditions. Thus, difficulties in obtaining permission to use genetically engineered microorganisms from government regulatory agencies as well as public controversies have made companies reluctant to develop bioremediation strategies based on GMOs. Finally, their use is considered costly. Technically speaking, it seems more plausible to use GMOs in ex-situ bioremediation treatment schemes in bioreactors, designed for use with defined soil slurries or water streams in tightly controlled environments. Not only does this limit the widespread release of the GMOs in the environment and avoids the problem of competition with indigenous microflora, but also allows the microorganism to be maintained at controlled temperatures and other growth conditions, to be used with relatively well-defined waste streams containing one or a small number of specific contaminants. The application of the genetically engineered microorganisms in industrial scale bioremediation is not yet prominent. Until today GMOs have not been used in commercial site remediation projects, with few only exceptions. Most bioaugmentation projects have used naturally-occurring bacteria for which obtaining regulatory approval is relatively easy. However, recently transgenic plants begin to find applicability in commercial phytoremediation projects.

Some common microbial constraints in bioremediation include:

i. *Diversity of Microorganisms*: The success of bioremediation relies on the diversity of microorganisms available in the contaminated site. Different pollutants require different metabolic pathways for degradation. Therefore, a diverse microbial community increases the likelihood of finding microorganisms capable of degrading specific pollutants.

Metabolic Capabilities: Microorganisms possess a vast array of metabolic capabilities that enable them to break down various pollutants. These capabilities of microorganism are regulated by the various enzymatic activities. The enzyme may be either intracellular or extracellular depending upon the characteristics of the pollutants. If any substance is penetrate directly inside the cell then intracellular enzyme becomes responsible for their metabolization. While in case of polymer substances most of the enzymes act extracellular to break down the chemical structure. Some microbes can degrade complex organic compounds, while others can convert toxic substances into less harmful forms through processes like mineralization or detoxification.

ii. *Synergistic Interactions*: Many bioremediation processes rely on microbial communities composed of diverse species of bacteria, archaea, and fungi. These microorganisms can work together synergistically to degrade complex pollutants.

For example, in the biodegradation of petroleum hydrocarbons, some bacteria may break down large hydrocarbons into simpler compounds, which are then further metabolized by other microorganisms. Microbial consortia can also cooperate by providing essential nutrients or cofactors to one another, ensuring that all metabolic pathways required for biodegradation are functional.

Co-metabolism is a phenomenon where microorganisms transform pollutants as a byproduct of their normal metabolic activities. This often occurs when a microorganism is degrading a primary substrate but can also transform a co-occurring pollutant. For example, during the degradation of chlorinated solvents, certain bacteria can co-metabolically transform these compounds while using a different, more easily metabolizable substrate as their primary energy source. In some cases, sequential bioremediation approaches may be employed, where different stages of remediation are optimized for specific processes. For example, in situ bioremediation may be used initially to reduce contaminant concentrations, followed by ex situ bioremediation to further treat the extracted material. These sequential strategies can allow for a more comprehensive and efficient removal of contaminants by leveraging synergistic interactions at different stages of the remediation process.

iii. *Nutrient Limitations*: Microorganisms require essential nutrients, such as carbon, nitrogen, phosphorus, and trace elements, for their growth and metabolic activity. In contaminated sites, these nutrients may be scarce or imbalanced, limiting the microbial population's growth and pollutant degradation capacity. Supplementing the environment with appropriate nutrients can help overcome this constraint.

iv. *Oxygen Availability*: As mentioned earlier, many pollutant-degrading microorganisms are aerobic, meaning they require oxygen for respiration. In anaerobic or low-oxygen environments, the activity of these microorganisms can be restricted. Techniques such as aeration or the use of oxygen-releasing compounds can alleviate this constraint. Oxygen availability can enhance energy production, microbial activity, pollutant transformation, bioavailability of contaminants, preventing harmful byproducts, redox potential control, temperature regulation, microbial community diversity

v. *pH and Temperature*: Microbial activity is influenced by pH and temperature. Extremes in these factors can inhibit microbial growth and enzymatic activity. Bioremediation efforts must consider the optimal pH and temperature ranges for the specific microbial strains involved.

vi. *Adaptation and Evolution*: Microorganisms have the ability to adapt and evolve rapidly in response to changing environmental conditions. This adaptability is crucial in bioremediation because pollutant concentrations and environmental conditions can vary over time. The microbial community can potentially adjust and optimize its biodegradation capabilities accordingly. The alterations in the cytoplasmic membrane composition play crucial role in adaptation to the presence of high concentrations of toxic contaminants. The major changes during adaptation are as:

» Changes in the fatty acid composition of membrane lipids

- Isomerization of Unsaturated Fatty Acids
- Changes in Cyclopropane and Branched Fatty Acids

» Changes in Phospholipids

» Synthesis of Cardiolipin (diphosphatidylglycerol) : unique phospholipid that plays an important role in cell membrane adaptation

» Efflux Pumps and Solvent Transport

» Production of Special Proteins : Induction of stress proteins in E. coli with aromatic compounds, such as 2,4-dinitrophenol (DNP) and benzoate. Other stress proteins DnaK and GroEL are induced by 2,4-dichlorophenoxyacetic acid in *Burkholderia* sp. YK-2 and by 4-chlorobiphenyl and biphenyl in *B. xenovorans* LB40. Expression regulation of the stress proteins. The role of alternative sigma factor σ B in this adaptation was emphasized. This factor controls the production of bmrUR operon in G+ *Bacillus subtilis* necessary for the production of multidrug efflux proteins.

vii. Competition and Predation: Microbial competition can impact the effectiveness of bioremediation. In some cases, certain microorganisms may outcompete the pollutant-degrading microbes for resources, limiting the overall degradation rate. On the other hand, predatory microorganisms can also play a role by feeding on competing organisms and promoting the growth of pollutant degraders.

viii. Toxicity and Inhibitory Compounds: Some pollutants or their breakdown products can be toxic to certain microorganisms, inhibiting their growth and activity. Therefore, microbial constraints involve identifying and selecting microorganisms that are resistant to or capable of detoxifying such inhibitory compounds.

ix. Spatial Heterogeneity: Contaminated environments are often spatially heterogeneous, with variations in pollutant concentrations, oxygen levels, and other environmental factors. This can create microenvironments where certain microorganisms are more favorable than others, leading to patchy or incomplete biodegradation.

x. *Monitoring and Control*: Understanding the microbial constraints involved in bioremediation is crucial for effective monitoring and control of the process. Regular assessment of the microbial community's composition and activity ensures that the bioremediation efforts are on track and adjustments can be made if needed.

xi. *Genetic Limitations:* The genetic diversity and metabolic capabilities of microorganisms in the contaminated site may not be sufficient to effectively degrade certain complex or recalcitrant pollutants.

10.2. Chemical Constrains

i. ***Bioavailability of contaminants***

The contaminants must be bioavailable for biological degradation to occur. The physical condition of the contaminant and the potential for effective contact between the microbe and the pollutant are linked to bioavailability. When the microorganism-contaminant interface is maximized, this contact is optimal. In terms of physical state, microorganisms often absorb pollutants from the liquid phase and are unable to efficiently break down a pollutant until it dissolves from nonaqueous phase liquids (NAPLs) into the bulk solution, diffuses out of nanopores, or desorbs from aquifer solids (Rayu et al., 2012). In these situations, the rates of diffusion, desorption, or dissolution can regulate the rate of biodegradation. Pollutants that are polar and soluble in water are more readily accessible. For hydrophobic pollutants, the addition of surface active agents may be necessary due to the increase in the contaminant-microorganism interaction surface. Determining a chemical's bioavailability requires an understanding of its partitioning and transfer rates between its dissolved, sorbed, and volatile phases. The effects of all the physical and chemical factors that ultimately determine a compound's capability for microbial utilization and, consequently, its potential for biodegradation are collectively referred to as bioavailability.

ii. ***Biodegradability of contaminants***

The success of any bioremediation project depends mainly on the chemical structure of the organic molecules present in the degraded site (Neilson and Allard, 2008). Some structural features of organic compounds that are not common in nature, called xenophores "(e.g., substitutions of H with Cl, NO_2, CN, and SO_3 groups), make such molecules difficult to be metabolized by microorganisms. As a result, pollutants containing these xenophores are typically resistant to microbial breakdown. The biodegradation of a broad range of organic molecules has been explained by a number of mechanisms and pathways. Enzymes mediate every metabolic reaction. These are oxidoreductases, hydrolases, lyases, transferases, isomerases, and ligases. Because of their non-specific substrate preference, many oxygenase enzymes that target aromatic hydrocarbons have an exceptionally broad degradation capacity. For instance, more than 100 distinct substances, such as TCE, nitrobenzene, and chlorobenzene, can be broken down by toluene dioxygenase. Other examples are esterases, which break down ester bonds by the addition of water; depolymerases, which hydrolyze polymers; dehalogenases, which remove halogen atoms such as chlorine and replace them with -OH groups; decarboxylases which remove CO_2 groups (i.e., decarboxylation), hydratases which add water to alkenes converting them into secondary alcohols; glutathione S-transferase which transfers the thiol group tochlorinated compounds with concomitant dechlorination; racemases which catalyze L and D-amino acid interconversions and finally CoA-ligase, which adds -S-CoA to fatty acids during beta-oxidation.

iii. Properties of other contaminant

Contaminant properties are critical to contaminant-soil interactions, contaminant mobility and to the ability of treatment technologies to remove, destroy or immobilize contaminants. Important contaminant properties include: Solubility in water, dielectric constant, diffusion coefficient, molecular weight, vapor pressure, density and aqueous solution chemistry.

Nutrients: Most in-situ bioremediation methods practiced today rely on the stimulation of indigenous microbial populations at the site of contamination, by addition of appropriate nutrients, principally carbon, oxygen, nitrogen and phosphorus, and by maintaining optimum conditions of pH, moisture and other factors, to trigger increased growth and activity of indigenous biodegrading microorganisms. Nitrogen and phosphorus requirements are often estimated by calculating a carbon to nitrogen to phosphorus ratio C/N/P close to 100/(10 to 5)/1. Nutrients are supplied by regulating ground water flow with injection wells or burred perforated pipes (infiltration gallery). Typically, groundwater extracted from production wells down gradient from the biostimulation zone is treated to remove impurities, amended with the nutrients needed for biostimulation, and then reintroduced to the aquifer up gradient of the biostimulation zone using injection wells or infiltration galleries. If the flow of extracted water is not enough to regulate the subsurface flow, an external water source is needed. Therefore, the solubility of the nutrients in water and the reinjection flow rate frequently limit the rate of nutrient delivery to the biostimulation zone.

iv. Oxygen, air, hydrogen peroxide

Bioremediation is an oxidation process in the majority of applications. Microorganisms use electron transfer to obtain energy during the oxidation of pollutants. During aerobic biodegradation, electrons are transferred from the pollutant to a terminal electron acceptor, which is oxygen. As the organic substrate breaks down in the subsurface, oxygen concentrations may decrease. Because oxygen is poorly soluble in water, the primary kinetic constraint for aerobic bioremediation is the availability of oxygen. When it comes to organic molecules that require a lot of oxygen, such petroleum hydrocarbons, this is more severe. To encourage aerobic biodegradation, the infiltration water may be supplemented with air, oxygen, or alternative oxygen sources (such as hydrogen peroxide or ozone).

Oxygen, air, and hydrogen peroxide play critical roles in bioremediation processes as they significantly influence the metabolic activity of microorganisms involved in pollutant degradation. Let's explore their roles in bioremediation:

Oxygen (O_2):

» *Aerobic bioremediation*: Many microorganisms involved in bioremediation are aerobic, meaning they require oxygen to carry out their metabolic processes. These microorganisms use oxygen as a terminal electron acceptor in their respiration, breaking down organic compounds into simpler and less toxic forms (Abatenh et al., 2017). Common pollutants that can be degraded aerobically include petroleum hydrocarbons, chlorinated solvents, and certain pesticides.

» *Aeration*: In some bioremediation strategies, oxygen may be introduced into the contaminated site by aeration. Aeration promotes the growth and activity of aerobic microorganisms, enhancing their ability to degrade pollutants. This method is particularly useful in enhancing the efficiency of in-situ bioremediation, where the contaminated soil or groundwater is treated in place.

Air (Oxygen and Nitrogen):

» *In-situ bioremediation*: Air is often used in in-situ bioremediation techniques such as biopiles and landfarming. In biopiles, contaminated soil is heaped into piles and aerated to provide oxygen to the microorganisms responsible for degradation. Landfarming involves spreading contaminated soil over a designated area and aerating it to stimulate microbial activity. Both methods allow for controlled degradation of pollutants by optimizing the supply of oxygen and nutrients to the microbial community.

Hydrogen Peroxide (H_2O_2):

» *Enhanced Bioremediation*: Hydrogen peroxide is sometimes used as an oxygen source in enhanced bioremediation processes. When injected into the subsurface, hydrogen peroxide releases oxygen as it decomposes, creating an oxygen-rich environment. This can enhance the aerobic biodegradation of pollutants in anaerobic or low-oxygen environments.

» *Chemical Oxidation*: Additionally, hydrogen peroxide can be used in combination with other chemical oxidants in advanced oxidation processes to directly oxidize and degrade certain pollutants. This approach is particularly useful for recalcitrant compounds that are not easily biodegraded.

It's important to note that while oxygen, air, and hydrogen peroxide can be beneficial in promoting bioremediation; their application must be carefully controlled. Excessive oxygen supply can lead to the formation of harmful by-products and may inhibit the growth of anaerobic microorganisms responsible for the degradation of certain pollutants. Furthermore, improper handling of hydrogen peroxide can cause adverse effects on the environment and the microbial community.

v. Alternative electron acceptors

In bioremediation, electron acceptors play a crucial role in the microbial degradation of pollutants. Traditional electron acceptors, like oxygen (O_2), are commonly used in aerobic bioremediation. However, in environments where oxygen is limited or absent, microorganisms can utilize alternative electron acceptors to carry out anaerobic degradation. These alternative electron acceptors facilitate the breakdown of various pollutants into less harmful forms. Some of the key alternative electron acceptors used in bioremediation includes:

Nitrate (NO^{3-}): Under anaerobic conditions, certain microorganisms can use nitrate as an electron acceptor in a process known as denitrification. Denitrification converts nitrate to nitrogen gas (N_2), which is released into the atmosphere, and helps in the breakdown of organic pollutants.

Sulfate (SO_4^{2-}): Sulfate-reducing microorganisms are capable of using sulfate as an electron acceptor in the process of sulfate reduction. This leads to the formation of sulfides (e.g., hydrogen sulfide, H_2S) and promotes the degradation of various sulfur-containing compounds.

Carbon Dioxide (CO_2): Some microorganisms can carry out fermentation, a process where organic compounds are broken down into simpler products like carbon dioxide, ethanol, or other organic acids. Although CO_2 is not a pollutant, this process helps in the overall degradation of contaminants.

Iron (Fe^{3+}): Iron-reducing microorganisms use ferric iron (Fe^{3+}) as an electron acceptor in the process of iron reduction. This process converts Fe3+ to ferrous iron (Fe^{2+}), leading to the formation of iron sulfides, which can immobilize certain heavy metals and assist in their removal from the environment.

Manganese (Mn^{4+}): Similar to iron reduction, certain microorganisms can reduce manganese ions (Mn^{4+}) to manganese ions (Mn^{2+}), aiding in the removal of manganese and related pollutants.

Halogenated Compounds: In some cases, halogenated compounds (e.g., chlorinated solvents) themselves can act as electron acceptors. Certain microorganisms can utilize these compounds to carry out reductive dechlorination, breaking down the halogens and degrading the pollutants.

Sulfur Compounds: Sulfur-containing compounds, such as sulfates, sulfites, or elemental sulfur, can serve as alternative electron acceptors in specific bioremediation scenarios.

Methane (CH_4): Methane-producing microorganisms, known as methanogens, can utilize carbon dioxide and hydrogen as electron acceptors, producing methane in the process. While not directly involved in pollutant degradation, methanogenesis can be relevant in certain anaerobic environments.

vi. Metal ions:

Metal ions play a significant role in bioremediation, both as contaminants that require removal and as essential elements that can facilitate microbial degradation processes. The role of metal ions in bioremediation can be understood from two perspectives:

Contaminant Removal:

» *Heavy Metal Remediation*: Metal ions like lead (Pb), mercury (Hg), cadmium (Cd), chromium (Cr), and others are considered heavy metal pollutants that pose significant environmental risks. They can accumulate in soil and water, leading to toxicity and adverse effects on ecosystems and human health. Bioremediation techniques can be employed to remove these metal ions from contaminated sites.

» *Biosorption*: Certain microorganisms, as well as some biomaterials (e.g., bacterial biomass, algae, and fungal biomass), possess the ability to bind and accumulate heavy metal ions onto their cell surfaces. This process, known as biosorption, can effectively sequester metal ions from the environment, reducing their mobility and bioavailability.

» *Biomineralization*: Some microorganisms can precipitate heavy metal ions as metal sulfides or carbonates, forming stable and less toxic compounds. Biomineralization helps immobilize the metal ions, making them less mobile and reducing their potential for leaching into the environment.

Facilitating Microbial Degradation:

» *Enzyme Activation*: In some bioremediation processes, metal ions act as essential cofactors for microbial enzymes involved in the degradation of pollutants. For example, certain metalloenzymes require metal ions like iron (Fe) or manganese (Mn) as cofactors to function effectively in breaking down organic contaminants (Bhandari et al., 2021).

» *Redox Reactions*: Metal ions can participate in redox reactions, which are critical in biodegradation processes. They can act as electron acceptors or donors, facilitating the transfer of electrons during microbial metabolism. This is particularly relevant in anaerobic bioremediation, where metal ions like iron or sulfate serve as alternative electron acceptors for pollutant degradation.

» *Biostimulation*: Some bioremediation strategies involve the addition of metal ions or metal-containing compounds to the contaminated site. This process, known as biostimulation, aims to promote the growth and activity of specific microorganisms capable of degrading pollutants. Metal ions, such as iron or manganese, can act as nutrients, stimulating the activity of certain microbial groups involved in biodegradation.

vii. Toxic compounds

Even if the same chemicals are easily broken down at lower concentrations, high aqueous phase concentrations of particular pollutants can have harmful effects on microorganisms. In addition to preventing or slowing down microbial metabolic activity, toxicity frequently inhibits the formation of new biomass, which is necessary to promote the quick elimination of contaminants. The specific toxicants, their concentration, and the microorganisms exposed all affect the degree and mechanisms of toxicity. Certain chemical compounds can be harmful to microorganisms as well as targeted living forms like plants and insects. Herbicides, insecticides, fungicides, pesticides, and rodenticides are some of these substances. Furthermore, many microorganisms find some kinds of inorganic chemicals, such cyanides and azides, poisonous; but, after a time of microbial adaptation, these compounds may be broken down.

10.3. BOD/COD ratio

The COD/BOD ratio is an indicator for the presence of organic and inorganic matter present in the wastewater, compost and other natural and man-made material. COD is always greater than BOD; hence the ratio will be always greater than 1.

If BOD/COD is 0.92 to 1, waste water is fully biodegradable.

If BOD/COD > 0.6, biological treatment is acceptable.

If BOD/COD is between 0.3 to 0.6, then some adjustment is required.

If BOD/COD < 0.3, then the biological treatment is not possible.

10.4. Biogeochemical Parameters

Measurements of various biogeochemical parameters such as as dissolved oxygen (DO), redox potential, CO_2, and different nutrient and electron acceptor concentrations, including NH^{4+}, NO^{3-}, NO^{2-}, SO_4^{2-}, S^{2-}, and Fe^{2+}, can provide valuable insights into the existing microbial metabolic activity at the contaminated site. These parameters help assess the environmental conditions and the potential for natural or intrinsic bioremediation processes to occur. Here's how these measurements can indicate the microbial metabolic activity:

Dissolved Oxygen (DO): DO levels indicate the availability of oxygen in the environment. High DO levels are indicative of aerobic conditions, where aerobic microorganisms can thrive and utilize oxygen as an electron acceptor in pollutant degradation. Conversely, low DO levels suggest anaerobic or low-oxygen conditions, which may support different microbial groups involved in anaerobic degradation processes, such as denitrification, sulfate reduction, or methanogenesis.

CO_2: CO_2 levels can provide an indication of microbial respiration and metabolic activity. The production of CO_2 is a common outcome of microbial degradation processes, especially during aerobic respiration. Measuring CO2 levels can offer insights into the overall microbial activity at the site.

NH^{4+}, NO^{3-}, NO^{2-}, $SO_4{}^{2-}$, S^{2-}, and *Fe^{2+}* : These measurements are essential for understanding the availability of different nutrient and electron acceptor compounds in the environment. Ammonium (NH^{4+}), nitrate (NO^{3-}), nitrite (NO^{2-}), sulfate ($SO_4{}^{2-}$), sulfide (S^{2-}), and ferrous iron (Fe^{2+}) are used by various microbial groups as electron acceptors or donors in different redox reactions. The presence and concentrations of these compounds can indicate which microbial processes are likely to dominate at the site.

10.5. Environmental Factors

Environmental factors play a vital role in bioremediation, as they influence the activity and effectiveness of microorganisms involved in pollutant degradation. Understanding and optimizing these factors are crucial for successful bioremediation efforts (Alori et al., 2022). Some important environmental factors for bioremediation include:

pH: The pH level of the environment affects microbial activity and enzyme function. Different microbial strains have specific pH preferences for optimal growth and pollutant degradation. Controlling and adjusting the pH within the appropriate range can enhance bioremediation efficiency.

Temperature: Microbial activity is temperature-dependent, and different microbial strains have distinct temperature ranges for optimal growth and metabolism. Monitoring and maintaining the temperature at the suitable range for the specific microbial community is essential for effective bioremediation.

Oxygen (O_2) Levels: Oxygen availability is critical, particularly in aerobic bioremediation. Oxygen is used by aerobic microorganisms as an electron acceptor in pollutant degradation. Understanding the oxygen levels and ensuring adequate aeration can support aerobic microbial communities.

Redox Potential: The redox potential reflects the oxidation-reduction conditions in the environment. Positive redox potential values indicate oxidizing conditions, while negative values suggest reducing conditions. Monitoring redox potential helps identify the predominant electron acceptors and donors and guides the choice of bioremediation pathways.

Nutrient Concentrations: Essential nutrients like carbon, nitrogen, phosphorus, and trace elements are vital for microbial growth and metabolism. Evaluating nutrient concentrations helps identify potential nutrient limitations and provides the basis for adding necessary supplements to support microbial activity.

Mass transfer characteristics: Mass transport characteristics are used to calculate potential rates of movement of liquids or gases through soil and include: Soil texture, unsaturated hydraulic conductivity, dispersivity, moisture content vs. soil moisture tension, bulk density, porosity, hydraulic conductivity and infiltration rate. Site hydro geologic characteristics, Hydro geologic factors for consideration include aquifer type, hydraulic conductivity, hydro geologic gradient, permeability, recharge capability, depth to groundwater, moisture content/field capacity, thickness of the saturated zone, homogeneity, depth to contamination, extent of contamination, and plume stability.

Contaminant Concentrations: Accurate measurements of the pollutant concentrations at the site are essential for assessing the extent of contamination and tracking the progress of biodegradation over time.

Availability of Electron Acceptors: The presence and concentrations of electron acceptors (e.g., oxygen, nitrate, sulfate, iron) influence the type of microbial processes that can occur. Providing suitable electron acceptors can enhance pollutant degradation.

Evaluation of Toxicity at site: Evaluating the toxicity of the contaminated site is crucial, as high toxicity levels can negatively affect microbial communities and hinder biodegradation processes.

Salinity: High salinity can be detrimental to many microorganisms, especially those not adapted to saline environments. Saline conditions can disrupt cellular osmotic balance and hinder microbial activity.

Organic Matter: The presence of organic matter can serve as a carbon source for microorganisms and support their growth and pollutant degradation capabilities.

Spatial Heterogeneity: Contaminated sites are often spatially heterogeneous, with variations in pollutant concentrations and environmental conditions. This can influence the distribution and activity of pollutant-degrading microorganisms.

10.6. Conclusion

Bioremediation is a multidisciplinary process and its success depends upon monitoring of various parameters requires for the growth of microorganism involved in the process. Hence, the success of bioremediation is heavily influenced by a combination of factors that interact in complex ways. Understanding these factors is crucial for designing effective and efficient bioremediation strategies to tackle environmental contamination. The key takeaway is that bioremediation is not a one-size-fits-all approach, and each contaminated site requires a tailored solution. Microbial activity, driven by suitable microorganisms, is at the heart of bioremediation processes. Maintaining optimal environmental conditions, including temperature, pH, oxygen levels, and nutrient availability, is essential for promoting microbial growth and activity. Additionally, the type and properties of the contaminant play a significant role in determining the most appropriate bioremediation technique. The availability of necessary nutrients for microbial metabolism and the potential presence of inhibitory substances must be carefully considered. Furthermore, the accessibility of the contaminant and the site-specific distribution influence the effectiveness of bioremediation. Patience is required as bioremediation is generally a time-consuming process, and the timeframe for complete remediation can vary widely. Regular monitoring and adaptive management are crucial for assessing the progress and making necessary adjustments to enhance bioremediation effectiveness. While bioremediation offers a sustainable and eco-friendly solution, ongoing research and technological advancements are vital to expand its applicability and optimize its performance. By comprehensively evaluating and addressing the various factors influencing bioremediation, we can harness the power of nature's microorganisms to restore contaminated environments and pave the way for a cleaner and healthier planet.

References

Abatenh, E., Gizaw, B., Tsegaye, Z., & Wassie, M. (2017). The role of microorganisms in bioremediation-A review. *Open Journal of Environmental Biology*, *2*(1), 038-046.

Abbasi, B., Nasiri, V., Azarfar, B., & Riemenschnitter, K. (2021). Biomining thriving on life-mineral coevolution.

Adhikari, S., Mishra, A. K., & Sarkar, D. (2018). Ceramic-Based Nanomaterials for Multifunctional Application. In *Smart Ceramics* (pp. 73-120). Jenny Stanford Publishing.

Admassu, W., & Korus, R. A. (1996). Engineering of bioremediation processes: needs and limitations. *Biotechnology Research Series*, *6*, 13-34.

Alcalde, M. (2007). Laccases: biological functions, molecular structure and industrial applications. In *Industrial Enzymes: structure, function and applications* (pp. 461-476). Dordrecht: Springer Netherlands.

Alori, E. T., Gabasawa, A. I., Elenwo, C. E., & Agbeyegbe, O. O. (2022). Bioremediation techniques as affected by limiting factors in soil environment. *Frontiers in Soil Science*, *2*, 937186.

Angulo-Bejarano, P. I., Puente-Rivera, J., & Cruz-Ortega, R. (2021). Metal and metalloid toxicity in plants: An overview on molecular aspects. Plants, 10(4), 635.

Arevalo-Gallegos, A., Ahmad, Z., Asgher, M., Parra-Saldivar, R., & Iqbal, H. M. (2017). Lignocellulose: a sustainable material to produce value-added products with a zero waste approach—a review. *International journal of biological macromolecules*, *99*, 308-318.

Ayilara, M. S., & Babalola, O. O. (2023). Bioremediation of environmental wastes: the role of microorganisms. *Frontiers in Agronomy*, *5*, 1183691.

Aziz, K. H. H., Mustafa, F. S., Omer, K. M., Hama, S., Hamarawf, R. F., & Rahman, K. O. (2023). Heavy metal pollution in the aquatic environment: efficient and low-cost removal approaches to eliminate their toxicity: a review. *RSC advances*, *13*(26), 17595-17610.

Bao, Y., Dolfing, J., Wang, B., Chen, R., Huang, M., Li, Z., ... & Feng, Y. (2019). Bacterial communities involved directly or indirectly in the anaerobic degradation of cellulose. *Biology and Fertility of Soils*, *55*, 201-211.

Bell, C. H., Wong, J., Parsons, K., Semel, W., McDonough, J., & Gerber, K. (2022). First Full-Scale In Situ Propane Biosparging for Co-Metabolic Bioremediation of 1, 4-Dioxane. *Groundwater Monitoring & Remediation*, *42*(4), 54-66.

Bhandari, S., Poudel, D. K., Marahatha, R., Dawadi, S., Khadayat, K., Phuyal, S., ... & Parajuli,

N. (2021). Microbial enzymes used in bioremediation. *Journal of Chemistry, 2021*, 1-17.

Bharadwaj, A. (2018). Bioremediation of xenobiotics: an eco-friendly cleanup approach. In *Green Chemistry in Environmental Sustainability and Chemical Education: Proceedings of ICGC 2016, New Delhi* (pp. 1-13). Springer Singapore.

Bolan, N. S., Park, J. H., Robinson, B., Naidu, R., & Huh, K. Y. (2011). Phytostabilization: a green approach to contaminant containment. *Advances in agronomy, 112*, 145-204.

Bradley, K. (2019). *Reproduction in Bacteria*. Scientific e-Resources.

Brigham, J. S., Adney, W. S., & Himmel, M. E. (2018). Hemicellulases: diversity and applications. In *Handbook on bioethanol* (pp. 119-141). Routledge.

Celin, S. M., Sahai, S., Kalsi, A., & Bhanot, P. (2020). Environmental monitoring approaches used during bioremediation of soils contaminated with hazardous explosive chemicals. *Trends in environmental analytical chemistry, 26*, e00088.

Chen, D., Gao, A., Cen, K., Zhang, J., Cao, X., & Ma, Z. (2018). Investigation of biomass torrefaction based on three major components: Hemicellulose, cellulose, and lignin. *Energy conversion and management, 169*, 228-237.

Croom, E. (2012). Metabolism of xenobiotics of human environments. *Progress in molecular biology and translational science, 112*, 31-88.

Dangi, A. K., Sharma, B., Hill, R. T., & Shukla, P. (2019). Bioremediation through microbes: systems biology and metabolic engineering approach. *Critical reviews in biotechnology, 39*(1), 79-98.

Demir, E. (2021). A review on nanotoxicity and nanogenotoxicity of different shapes of nanomaterials. *Journal of Applied Toxicology, 41*(1), 118-147.

Di Giulio, R. T., Benson, W. H., Sanders, B. M., & Van Veld, P. A. (2020). Biochemical mechanisms: metabolism, adaptation, and toxicity. In *Fundamentals of aquatic toxicology* (pp. 523-561). CRC Press.

Díez-Pascual, A. M. (2021). Carbon-based nanomaterials. *International Journal of Molecular Sciences, 22*(14), 7726.

Dutta, J., & Choudhury, M. (2018). Plastic pollution: a global problem from a local perspective. *Journal of Waste Management & Xenobiotics, 1*(1), 000102.

Farraji, H., Zaman, N. Q., Tajuddin, R., & Faraji, H. (2016). Advantages and disadvantages of phytoremediation: A concise review. *International Journal of Environmental Science and Technology, 2*, 69-75.

Filler, D. M., Snape, I., & Barnes, D. L. (Eds.). (2008). *Bioremediation of petroleum hydrocarbons in cold regions*. Cambridge University Press.

Fu, Z., & Xi, S. (2020). The effects of heavy metals on human metabolism. *Toxicology mechanisms and methods, 30*(3), 167-176.

Gandhi, M., Moghal, A. A. B., Rasheed, R. M., & Almajed, A. (2022). State-of-the-art review on geoenvironmental benign applicability of biopiles. *Innovative Infrastructure Solutions*, *7*(2), 166.

Gavrilescu, M. (2022). Enhancing phytoremediation of soils polluted with heavy metals. *Current Opinion in biotechnology*, *74*, 21-31.

Glasser, W. G., Atalla, R. H., Blackwell, J., Malcolm Brown, R., Burchard, W., French, A. D., ... & Nishiyama, Y. (2012). About the structure of cellulose: debating the Lindman hypothesis. *Cellulose*, *19*, 589-598.

Gohil, H., & Ogram, A. (2020). The microbial degradation of DDT and potential remediation strategies. *Bioremediation Technology*, 1-24.

Gold, M. H., Youngs, H. L., & Gelpke, M. D. S. (2000). Manganese peroxidase. *Metal ions in biological systems*, 607-634.

Goldsbrough, P. (2020). Metal tolerance in plants: the role of phytochelatins and metallothioneins. In *Phytoremediation of contaminated soil and water* (pp. 221-233). Crc Press.

Han, J., Zhao, D., Li, D., Wang, X., Jin, Z., & Zhao, K. (2018). Polymer-based nanomaterials and applications for vaccines and drugs. *Polymers*, *10*(1), 31.

Hashmi, M. Z., Kumar, V., & Varma, A. (Eds.). (2017). *Xenobiotics in the soil environment: monitoring, toxicity and management* (Vol. 49). Springer.

Houfani, A. A., Anders, N., Spiess, A. C., Baldrian, P., & Benallaoua, S. (2020). Insights from enzymatic degradation of cellulose and hemicellulose to fermentable sugars–a review. *Biomass and Bioenergy*, *134*, 105481.

Jadaun, J. S., Bansal, S., Sonthalia, A., Rai, A. K., & Singh, S. P. (2022). Biodegradation of plastics for sustainable environment. *Bioresource Technology*, *347*, 126697.

Jannasch, H. W., & Egli, T. (1993). Microbial growth kinetics: a historical perspective. *Antonie van Leeuwenhoek*, *63*, 213-224.

Jayaramaiah, R. H., Egidi, E., Macdonald, C. A., Wang, J. T., Jeffries, T. C., Megharaj, M., & Singh, B. K. (2022). Soil initial bacterial diversity and nutrient availability determine the rate of xenobiotic biodegradation. *Microbial Biotechnology*, *15*(1), 318-336.

Jørgensen, K. S., Puustinen, J., & Suortti, A. M. (2000). Bioremediation of petroleum hydrocarbon-contaminated soil by composting in biopiles. *Environmental pollution*, *107*(2), 245-254.

Kamimura, N., Sakamoto, S., Mitsuda, N., Masai, E., & Kajita, S. (2019). Advances in microbial lignin degradation and its applications. *Current Opinion in Biotechnology*, *56*, 179-186.

Kaur, S., Samota, M. K., Choudhary, M., Choudhary, M., Pandey, A. K., Sharma, A., & Thakur, J. (2022). How do plants defend themselves against pathogens-Biochemical mechanisms and genetic interventions. *Physiology and Molecular Biology of Plants*, *28*(2), 485-504.

Khanpour-Alikelayeh, E., & Partovinia, A. (2021). Synergistic and antagonistic effects of microbial co-culture on bioremediation of polluted environments. *Microbial Rejuvenation of Polluted Environment: Volume 2*, 229-265.

Laurila, T., Sainio, S., & Caro, M. A. (2017). Hybrid carbon based nanomaterials for electrochemical detection of biomolecules. *Progress in Materials Science*, *88*, 499-594.

Li, M., & Wen, J. (2021). Recent progress in the application of omics technologies in the study of bio-mining microorganisms from extreme environments. *Microbial Cell Factories*, *20*(1), 178.

Limmer, M., & Burken, J. (2016). Phytovolatilization of organic contaminants. *Environmental Science & Technology*, *50*(13), 6632-6643.

Liu, L. Y., Hua, Q., & Renneckar, S. (2019). A simple route to synthesize esterified lignin derivatives. *Green chemistry*, *21*(13), 3682-3692.

Liu, R., Zhao, Y., Doherty, L., Hu, Y., & Hao, X. (2015). A review of incorporation of constructed wetland with other treatment processes. *Chemical Engineering Journal*, *279*, 220-230.

Luo, Y., Li, Z., Li, X., Liu, X., Fan, J., Clark, J. H., & Hu, C. (2019). The production of furfural directly from hemicellulose in lignocellulosic biomass: A review. *Catalysis Today*, *319*, 14-24.

Ma, H., Zhao, Y., Yang, K., Wang, Y., Zhang, C., & Ji, M. (2022). Application oriented bioaugmentation processes: Mechanism, performance improvement and scale-up. *Bioresource Technology*, *344*, 126192.

Madadi, M., & Abbas, A. (2017). Lignin degradation by fungal pretreatment: a review. *J Plant Pathol Microbiol*, *8*(2), 1-6.

Magdziak, Z., Gąsecka, M., Goliński, P., & Mleczek, M. (2015). Phytoremediation and environmental factors. *Phytoremediation: Management of Environmental Contaminants, Volume 1*, 45-55.

Maier, R. M., & Pepper, I. L. (2015). Bacterial growth. In *Environmental microbiology* (pp. 37-56). Academic Press.

Maila, M. P., & Cloete, T. E. (2004). Bioremediation of petroleum hydrocarbons through landfarming: are simplicity and cost-effectiveness the only advantages?. *Reviews in Environmental science and bio/Technology*, *3*, 349-360.

Mathews, S. L., Pawlak, J., & Grunden, A. M. (2015). Bacterial biodegradation and bioconversion of industrial lignocellulosic streams. *Applied microbiology and biotechnology*, *99*, 2939-2954.

Meshitsuka, G., & Isogai, A. (2017). Chemical structures of cellulose, hemicelluloses, and lignin. In *Chemical modification of lignocellulosic materials* (pp. 11-33). Routledge.

Michelin, M., Ruíz, H. A., Silva, D. P. D., Ruzene, D. S., Teixeira, J. A., & Polizeli, M. D. L. (2015). Cellulose from lignocellulosic waste.

Miglani, R., Parveen, N., Kumar, A., Ansari, M. A., Khanna, S., Rawat, G., ... & Ansari, M. N. (2022). Degradation of xenobiotic pollutants: an environmentally sustainable approach. *Metabolites*, *12*(9), 818.

Mishra, M., Singh, S. K., & Kumar, A. (2021). Environmental factors affecting the bioremediation potential of microbes. In *Microbe Mediated Remediation of Environmental Contaminants* (pp. 47-58). Woodhead Publishing.

Mohapatra, B., & Phale, P. S. (2021). Microbial degradation of naphthalene and substituted naphthalenes: metabolic diversity and genomic insight for bioremediation. *Frontiers in Bioengineering and Biotechnology*, *9*, 602445.

Moreno, R., & Rojo, F. (2023). The importance of understanding the regulation of bacterial metabolism. *Environmental microbiology*, *25*(1), 54.

Muter, O. (2023). Current trends in bioaugmentation tools for bioremediation: A critical review of advances and knowledge gaps. *Microorganisms*, *11*(3), 710.

Narayanan, M., Ali, S. S., & El-Sheekh, M. (2023). A comprehensive review on the potential of microbial enzymes in multipollutant bioremediation: Mechanisms, challenges, and future prospects. *Journal of Environmental Management*, *334*, 117532.

Nasreen, Z., & Kalsoom, S. (2018). Biodegradation of petroleum industry oily sludge and its application in land farming: a review. *International Journal of Environment and Waste Management*, *21*(1), 37-57.

Newman, L. A., & Reynolds, C. M. (2004). Phytodegradation of organic compounds. *Current opinion in Biotechnology*, *15*(3), 225-230.

Newsome, L., & Falagán, C. (2021). The microbiology of metal mine waste: Bioremediation applications and implications for planetary health. *GeoHealth*, *5*(10), e2020GH000380.

Pande, V., Pandey, S. C., Sati, D., Pande, V., & Samant, M. (2020). Bioremediation: an emerging effective approach towards environment restoration. *Environmental Sustainability*, *3*, 91-103.

Panigrahi, S., Velraj, P., & Rao, T. S. (2019). Functional microbial diversity in contaminated environment and application in bioremediation. In *Microbial diversity in the genomic era* (pp. 359-385). Academic press.

Parde, D., Patwa, A., Shukla, A., Vijay, R., Killedar, D. J., & Kumar, R. (2021). A review of constructed wetland on type, treatment and technology of wastewater. *Environmental Technology & Innovation*, *21*, 101261.

Piontek, K., Smith, A. T., & Blodig, W. (2001). Lignin peroxidase structure and function. *Biochemical Society Transactions*, *29*(2), 111-116.

Piwowarska, D., & Kiedrzyńska, E. (2022). Xenobiotics as a contemporary threat to surface waters. *Ecohydrology & Hydrobiology*, *22*(2), 337-354.

Qin, W., Wang, C. Y., Ma, Y. X., Shen, M. J., Li, J., Jiao, K., ... & Niu, L. N. (2020). Microbe-mediated extracellular and intracellular mineralization: environmental, industrial, and biotechnological applications. *Advanced Materials, 32*(22), 1907833.

Qiu, G., Liu, X., & Zhang, R. (2022). Biomining in China: History and Current Status. In *Biomining Technologies: Extracting and Recovering Metals from Ores and Wastes* (pp. 151-161). Cham: Springer International Publishing.

Rayu, S., Karpouzas, D. G., & Singh, B. K. (2012). Emerging technologies in bioremediation: constraints and opportunities. *Biodegradation, 23*, 917-926.

Robinson, B. H., Anderson, C. W. N., & Dickinson, N. M. (2015). Phytoextraction: where's the action?. *Journal of Geochemical Exploration, 151*, 34-40.

Robles-González, I. V., Fava, F., & Poggi-Varaldo, H. M. (2008). A review on slurry bioreactors for bioremediation of soils and sediments. *Microbial Cell Factories, 7*, 1-16.

Romantschuk, M., Lahti-Leikas, K., Kontro, M., Galitskaya, P., Talvenmäki, H., Simpanen, S., ... & Sinkkonen, A. (2023). Bioremediation of contaminated soil and groundwater by in situ biostimulation. *Frontiers in Microbiology, 14*, 1258148.

Rosa, E., Sufardi, S., Syafruddin, S., & Rusdi, M. (2023). Bioremediation of Ex-Mining Soil with the Biocompost in the Incubation Experiments. *Applied and Environmental Soil Science, 2023.*

Sahni, A., & Pei, D. (2021). Bacterial toxins escape the endosome by inducing vesicle budding and collapse. *ACS chemical biology, 16*(11), 2415-2422.

Saleh, T. A. (2020). Nanomaterials: Classification, properties, and environmental toxicities. *Environmental Technology & Innovation, 20*, 101067.

Saravanan, A., Kumar, P. S., Vo, D. V. N., Jeevanantham, S., Karishma, S., & Yaashikaa, P. R. (2021). A review on catalytic-enzyme degradation of toxic environmental pollutants: Microbial enzymes. *Journal of Hazardous Materials, 419*, 126451.

Sharma, J. (2019). Advantages and limitations of in situ methods of bioremediation. *Recent Adv Biol Med, 5*(2019), 10941.

Sharma, R. K., & Arora, D. S. (2015). Fungal degradation of lignocellulosic residues: an aspect of improved nutritive quality. *Critical reviews in microbiology, 41*(1), 52-60.

Shen, X., Dai, M., Yang, J., Sun, L., Tan, X., Peng, C., ... & Naz, I. (2022). A critical review on the phytoremediation of heavy metals from environment: Performance and challenges. *Chemosphere, 291*, 132979.

Singh, A., Prasad, S. M., & Singh, R. P. (Eds.). (2016). *Plant responses to xenobiotics* (Vol. 362). Singapore: Springer.

Smith, J. J., Gaito, S. T., & Koons, B. W. (2023). Bioventing revisited: efficacy of enhanced biodegradation for sites with mobile LNAPL. *Quarterly Journal of Engineering Geology and Hydrogeology, 56*(2), qjegh2022-085.

Sravya, K., & Sangeetha, S. (2022). Feasibility study on bioremediation techniques to contaminated soils. *Materials Today: Proceedings, 51*, 2556-2560.

Sreedevi, P. R., Suresh, K., & Jiang, G. (2022). Bacterial bioremediation of heavy metals in wastewater: a review of processes and applications. *Journal of Water Process Engineering, 48*, 102884.

Srivastava, J., Naraian, R., Kalra, S. J. S., & Chandra, H. (2014). Advances in microbial bioremediation and the factors influencing the process. *International Journal of environmental science and technology, 11*, 1787-1800.

Stentiford, E. I. (1996). Composting control: principles and practice. *The science of composting*, 49-59.

Tchounwou, P. B., Yedjou, C. G., Patlolla, A. K., & Sutton, D. J. (2012). Heavy metal toxicity and the environment. *Molecular, clinical and environmental toxicology: volume 3: environmental toxicology*, 133-164.

Tegene, B. G. (2020). Mode of action, mechanism and role of microbes in bioremediation service for environmental pollution management. *Journal of Biotechnology & Bioinformatics Research, 2*(3), 1-18.

Tekere, M., Jacob-Lopes, E., & Zepka, L. Q. (2019). Microbial bioremediation and different bioreactors designs applied. *Biotechnology and bioengineering*, 1-19.

Tripathi, S., Sanjeevi, R., Anuradha, J., Chauhan, D. S., & Rathoure, A. K. (2022). Nano-bioremediation: nanotechnology and bioremediation. In *Research anthology on emerging techniques in environmental remediation* (pp. 135-149). IGI Global.

Tsao, D. T. (2003). Overview of phytotechnologies. *Phytoremediation*, 1-50.

Tyagi, B., & Kumar, N. (2021). Bioremediation: Principles and applications in environmental management. In *Bioremediation for environmental sustainability* (pp. 3-28). Elsevier.

Ukaogo, P. O., Ewuzie, U., & Onwuka, C. V. (2020). Environmental pollution: causes, effects, and the remedies. In *Microorganisms for sustainable environment and health* (pp. 419-429). Elsevier.

Verma, S., & Kuila, A. (2019). Bioremediation of heavy metals by microbial process. *Environmental Technology & Innovation, 14*, 100369.

Vieira, V. H. A. D. M., & Matheus, D. R. (2019). Environmental assessments of biological treatments of biowaste in life cycle perspective: A critical review. *Waste Management & Research, 37*(12), 1183-1198.

Wang, H., Zhang, S., Pratush, A., Ye, X., Xie, J., Wei, H., ... & Hu, Z. (2018). Acclimation of culturable bacterial communities under the stresses of different organic compounds. *Frontiers in Microbiology, 9*, 225.

Wang, M., Chen, S., Jia, X., & Chen, L. (2021). Concept and types of bioremediation. In *Handbook of bioremediation* (pp. 3-8). Academic Press.

Wei, Z., Van Le, Q., Peng, W., Yang, Y., Yang, H., Gu, H., ... & Sonne, C. (2021). A review on phytoremediation of contaminants in air, water and soil. *Journal of hazardous materials*, *403*, 123658.

Weimer, P. J. (2022). Degradation of cellulose and hemicellulose by ruminal microorganisms. *Microorganisms*, *10*(12), 2345.

Wong-Pinto, L. S., Menzies, A., & Ordóñez, J. I. (2020). Bionanomining: biotechnological synthesis of metal nanoparticles from mining waste—opportunity for sustainable management of mining environmental liabilities. *Applied microbiology and biotechnology*, *104*(5), 1859-1869.

Wu, Q., Miao, W. S., Zhang, Y. D., Gao, H. J., & Hui, D. (2020). Mechanical properties of nanomaterials: A review. *Nanotechnology Reviews*, *9*(1), 259-273.

Wu, W. M., Yang, J., & Criddle, C. S. (2017). Microplastics pollution and reduction strategies. *Frontiers of Environmental Science & Engineering*, *11*, 1-4.

For Product Safety Concerns and Information please contact our EU representative GPSR@taylorandfrancis.com Taylor & Francis Verlag GmbH, Kaufingerstraße 24, 80331 München, Germany